Phosphate and Vitamin D in Chronic Kidney Disease

Contributions to Nephrology

Vol. 180

Series Editor

Claudio Ronco Vicenza

Phosphate and Vitamin D in Chronic Kidney Disease

Volume Editor

Mohammed S. Razzaque Boston, Mass.

25 figures and 4 tables, 2013

KARGER Basel · Freiburg · Paris · London · New York · New Delhi · Bangkok · Beijing · Tokyo · Kuala Lumpur · Singapore · Sydney

Contributions to Nephrology
(Founded 1975 by Geoffrey M. Berlyne)

Mohammed S. Razzaque
Department of Oral Medicine, Infections & Immunity
Harvard School of Dental Medicine
190 Longwood Avenue
Research & Education Building, Room 304
Boston, MA 02115 (USA)

Library of Congress Cataloging-in-Publication Data

Phosphate and vitamin D in chronic kidney disease / volume editor, Mohammed S. Razzaque.
p. ; cm. -- (Contributions to nephrology, ISSN 0302-5144 ; vol. 180)
Includes bibliographical references and indexes.
ISBN 978-3-318-02369-5 (hard cover : alk. paper) -- ISBN 978-3-318-02370-1 (e-ISBN)
I. Razzaque, Mohammed S., editor of compilation. II. Series: Contributions to nephrology ; v. 180. 0302-5144
[DNLM: 1. Renal Insufficiency, Chronic--physiopathology. 2. Disease Progression. 3. Phosphates--metabolism. 4. Vitamin D--metabolism. W1 CO778UN v.180 2013 / WJ 342]
RC918.R4
616.6'14--dc23

2013010350

Bibliographic Indices. This publication is listed in bibliographic services, including Current Contents® and Index Medicus.

www.karger.com
Printed in Germany on acid-free and non-aging paper (ISO 9706) by Bosch Druck, Ergolding
ISSN 0302–5144
e-ISSN 1662–2782
ISBN 978–3–318–02369–5
e-ISBN 978–3–318–02370–1

Contents

Preface

The common thread connecting the chapters of this book is the two most affected nutrients in chronic kidney disease – phosphate and vitamin D. Chapters are committed to provide the physiological aspects of phosphate and vitamin D metabolism, and how their pathological dysregulation facilitates the advancement of chronic kidney disease and influences the eventual outcome of the patients. I selected the chapters to present a framework of what the readers would be able to follow to understand the complex molecular and organ cross-talks during disease progression, ranging from the involvement of the fibroblast growth factor 23-klotho system in impaired phosphate transport to parathyroid dysfunction. Each chapter clearly presents the clinically important problems and lists the key unsolved issues related to chronic kidney disease. Readers will be able to use this simple book as a quick reference for updated information on phosphate and vitamin D metabolism without prior acquaintance with the field. My time and efforts of writing, editing and organizing this book will prove worthwhile if its contents inspire young physicians and scientists to take on the challenge of solving some of these clinically imperative issues, through their dynamic thinking and innovative research works. Finally, I would like to express my thanks to the research scientists in my laboratory at the Harvard School of Dental Medicine in Boston who, through their hard work, generated an enormous amount of clinically and biologically meaningful data on phosphate and vitamin D metabolism in the shortest possible time. Lastly, this project would not have reached its present form without the help and continued encouragement from the Razzaque family members (Rafi, Yuki, Newaz, Zahid, Shahed, Lisa, Muhit and Nilufar Begum).

Mohammed S. Razzaque, Boston, Mass.

Razzaque MS (ed): Phosphate and Vitamin D in Chronic Kidney Disease.
Contrib Nephrol. Basel, Karger, 2013, vol 180, pp 1–13 (DOI: 10.1159/000346774)

Osteo-Renal Cross-Talk and Phosphate Metabolism by the FGF23-Klotho System

Mutsuko Ohnishi · Mohammed S. Razzaque

Department of Oral Medicine, Infection and Immunity, Harvard School of Dental Medicine, Boston, Mass., USA

Abstract

Phosphate is widely distributed in the body and an adequate balance is required for maintaining essential cellular and organ functions. Dysregulation of phosphate balance, either in the form of hypophosphatemia or hyperphosphatemia can induce disorders ranging from rickets/osteomalacia to cardiovascular calcification. A physiologic phosphate balance is delicately maintained by multiorgan cross-talks among the intestine, kidney, and bone. Sodium-dependent phosphate (Na/Pi) cotransporters present in the intestine and kidney play a major role in phosphate absorption and reabsorption, according to the body's demand. Some of the calcium regulating factors, including parathyroid hormone and vitamin D can influence the activities of Na/Pi cotransporters, and thereby can affect phosphate balance. In addition, molecular analysis of the unexplained hypophosphatemic diseases, including autosomal-dominant hypophosphatemic rickets and tumor-induced osteomalacia has led to the identification of fibroblast growth factor 23 (FGF23). Subsequent studies have documented that bone-derived FGF23 and kidney-derived klotho can form an endocrine network to control urinary phosphate excretion. Studies have also documented negative effect of FGF23/klotho system on vitamin D metabolism and Na/Pi cotransporter activities. This article will summarize how the FGF23/klotho system might influence systemic phosphate metabolism, and consequences of its abnormal regulation will be briefly described.

Adequate phosphate balance is essential for the maintenance of fundamental cellular functions of the mammalian system, ranging from energy metabolism to mineral ion metabolism [1]. Despite such important biological functions, the molecular regulation of phosphate metabolism is not clearly understood. One of the reasons for such a lack of understanding of phosphate metabolism lies in the fact that it was often considered to be a secondary process, and controlled by the calcium regulating factors, such as parathyroid hormone (PTH) and vitamin D. However, molecular analysis of certain hypophosphatemic and hyperphosphatemic diseases has established a calcium-independent active regulation of phosphate metabolism [2–6].

Phosphate is a widely distributed element in the body, with bone and teeth consuming more than 80% of total phosphate where it is stored as apatite. Various soft tissues, including skeletal muscle utilize remaining phosphate, while less than 0.1% of phosphate is present within the cells and in the extracellular fluids [1, 7–9]. Of particular importance, oxidative phosphorylation reaction uses intracellular phosphate ions to generate energy by forming adenosine triphosphate (ATP). More than 2,000 biochemical reactions within the cell need phosphate to effectively execute cellular machineries, and an imbalance of intra- and extracellular phosphate concentration can disrupt some of these essential cellular reactions. Of relevance, the ionic phosphate particles do not have high lipid solubility, and therefore cannot easily cross the cell-membrane barrier, and need a cell membrane transport protein to carry the ionic phosphate. The negatively charged ionic form of phosphate can circulate within the body, as free phosphorus, and constitutes 50% of such inorganic form. The rest of the phosphorus ion in the blood exists as HPO_4^{2-} and can form a complex with sodium, calcium, magnesium, etc. [10–13]. The ionic form of phosphate contributes to the maintenance of blood P^H, to preserve acid-base balance, according to the metabolic and respiratory need of the body. Such a buffering effect of blood P^H is achieved by shifting HPO_4^{2-} to HPO_4^- and consequently promoting the renal excretion of the hydrogen ions. In addition to the renal regulation of phosphate metabolism, several other organs, including intestine and bone are also actively involved in phosphate metabolism [1, 14–16].

Entero-Renal Regulation of Phosphate Metabolism

The physiologic balance of phosphate is delicately maintained by multiorgan cross-talks, mostly among the intestine, kidney and bone [14, 17]. Sodium-dependent phosphate (Na/Pi) cotransporters of the intestine and kidney play a major role in entero-renal regulation of phosphate metabolism, by influencing the rate of phosphate absorption and reabsorption. Phosphate absorption mostly occurs in the small intestine [17], where intestinal phosphate transporter, Na/Pi-IIb, helps in the uptake process. Dietary phosphate and 1,25-dihydroxyvitamin D, by influencing the activity of Na/Pi-IIb transporters, can affect the rate of intestinal phosphate absorption [18]. Of relevance, 1,25-dihydroxyvitamin D can induce the expression of intestinal Na/Pi-IIb protein to augment phosphate absorption. In a similar line of observation, the intestinal phosphate uptake in *Na/Pi-IIb* knockout mice, following acute administration of phosphate, has shown to be reduced by 50%, as compared to the wild-type mice [19]. It does, however, need to be mentioned that the loss-of-function mutations of the human Na/Pi-IIb gene did not significantly alter phosphate balance [20].

The kidney reabsorbs more than 80% of the filtered phosphate in the proximal tubular epithelial cells, with a minute amount reabsorbed in the distal convoluted tubules. Similar to the sodium-dependent intestinal absorption of phosphate, the renal reabsorption of phosphate is also a sodium-dependent process, mostly mediated by Na/Pi-IIa and Na/Pi-IIc transporters, located in the brush-border membrane of the renal proximal tubular epithelial cells [21–23]. The importance of type II receptors in renal phosphate reabsorption is evident from the identification of loss-of-function mutations of the human Na/Pi-IIc gene that resulted in an increased amount of urinary phosphate wasting in patients with hereditary hypophosphaturic rickets with hypercalciuria [24]. A similar phosphate-wasting phenotype has also been documented in mice lacking either Na/Pi-IIa or Na/Pi-IIc transporter activity [25, 26]. Of relevance, Na/Pi-IIa and Na/Pi-IIc transporters are believed to contribute to up to 70% of the phosphate reabsorption process.

In addition to Na/Pi type II transporters, the kidney also contains type I and type III transporters [27, 28], although the functions of these transporters in phosphate metabolism are not yet as clear as type II transporters. The type I transporters expressed in the liver and the kidney have low a affinity for phosphate compounds and are actively involved in the excretion of organic anions (uric acid, etc.) [29]. In contrast to the type I transporters, type III transporters (Pit-1 and Pit-2) are expressed more diffusely in various organs and have a higher affinity for phosphate [30, 31]. However, additional experimental validations are needed to establish the function of type III transporters as a vehicle for cellular phosphate translocation. Studies have also identified additional factors that might influence phosphate transport process, and in the accompanying chapter Caldas et al. [32] described one such potential with nuclear receptor LXR.

A numbers of factors, including PTH can control the activity of renal Na/Pi type II transporters, and thereby can influence systemic phosphate metabolism. PTH through binding with its receptor on the basolateral surface of the tubular epithelial cells can induce endocytotic internalization of Na/Pi-IIa and Na/Pi-IIc transporters to reduce phosphate reabsorption [27, 28]. The detail molecular regulations of Na/Pi system in phosphate metabolism are elaborated in the accompanying low, and whether such reduced levels of klotho might affect diseases progression is partly explained in the accompanying chapter by Hu et al. [88].

Maintaining serum phosphate levels within the normal range is a key strategy to effectively treat CKD patients, and to reduce or delay associated complications such as vascular calcification or mineralization; CKD patients are routinely advised to consume a low-phosphate-containing diet, and are often prescribed with phosphate binders to reduce intestinal phosphate intake. Unfortunately, lack of awareness of phosphate-containing foods and drinks has minimized the desired beneficial effects [89]. Almost 50% of mortality in the advanced stages of CKD patients is related to cardiovascular anomalies. In a similar line of study, hyperphosphatemia in CKD patients receiving dialysis treatment correlates with the incidence of vascular mineralization; more importantly, reducing the phosphate burden in these patients can attenuate progression of vascular mineralization [90]. The possible mechanism of phosphate-induced vascular mineralization is detailed in the accompanying chapter [91]. Studies have shown that a higher extracellular concentration of phosphate can induce vascular smooth muscle cell death, and vascular smooth muscle cell transdifferentiation into osteogenic cells to facilitate ectopic vascular mineralization. The accompanying chapters have discussed how extracellular phosphate can directly induce a subcellular signaling network [92], and the consequences of elevated extracellular phosphate concentration on eventual tissue events [93]; these observations raise the possibility of far-reaching toxic effects of phosphate beyond the abnormal mineralization of bone and soft tissues in patients with CKD [94].

Concluding Remarks

The loss of kidney function in patients with CKD disrupts normal mineral ion balance, as is evident from the abnormal serum levels of calcium, phosphate, PTH and vitamin D; such dysregulation adversely affects bone remodeling and facilitates formation of ectopic calcification (fig. 1). Since advanced stages of CKD affect multiple systems and organs, it is often challenging to manage these patients and requires continuous long-term monitoring. Efforts to reduce the phosphate burden by using phosphate binders have limited success due to associated complications. Calcitriol and other new vitamin D analogs are often tried to treat secondary hyperparathyroidism in CKD patients, while calcimimetic agents (cinacalcet hydrochloride) are also prescribed to reduce PTH levels. In some extreme uncontrolled cases of CKD, parathyroidectomy is recommended. It is, however, necessary to mention that despite the availability of new-generation drugs, the overall mortality of CKD patients is not dramatically improved, implicating a need for further exploration to find and produce more effective drugs. Of relevance, both vitamin D and PTH dysregulation in patients with CKD is the consequence of abnormal phosphate metabolism due to osteorenal miscommunication. Developing strategies for better control of phosphate balance, perhaps by manipulating the FGF23/klotho system, in the earlier stages of CKD are likely to minimize nonrenal complications to reduce overall morbidity and mortality of the patients.

Acknowledgements

Part of the research work is supported by the grant (R01-DK077276 to M.S.R.) provided by the NIH (NIDDK). Thanks are due to the members of Razzaque lab, including, Teruyo Nakatani (PhD), Shigeko Kato (PhD), Junko Akiyoshi (MD), Satoko Osuka (MD), Yongguen Hong (PhD), and Khadijah Turkistani (BDS) of Harvard School of Dental Medicine (Boston) for their original research works, some of which formed the basis of this book chapter.

References

1 Gaasbeek A, Meinders AE: Hypophosphatemia: an update on its etiology and treatment. Am J Med 2005;118:1094–1101.

2 ADHR_Consortium: Autosomal dominant hypophosphataemic rickets is associated with mutations in FGF23. The ADHR consortium. Nat Genet 2000;26:345–348.

3 Tenenhouse HS: X-linked hypophosphataemia: a homologous disorder in humans and mice. Nephrol Dial Transplant 1999;14:333–341.

4 Chong WH, Molinolo AA, Chen CC, Collins MT: Tumor-induced osteomalacia. Endocr Relat Cancer 2011;18:R53–R77.

5 Endo I, Fukumoto S, Ozono K, Namba N, Tanaka H, Inoue D, Minagawa M, Sugimoto T, Yamauchi M, Michigami T, Matsumoto T: Clinical usefulness of measurement of fibroblast growth factor 23 (FGF23) in hypophosphatemic patients: proposal of diagnostic criteria using FGF23 measurement. Bone 2008; 42:1235–1239.

6 Shimada T, Mizutani S, Muto T, Yoneya T, Hino R, Takeda S, Takeuchi Y, Fujita T, Fukumoto S, Yamashita T: Cloning and characterization of FGF23 as a causative factor of tumor-induced osteomalacia. Proc Natl Acad Sci USA 2001;98:6500–6505.

7 Razzaque MS: FGF23-mediated regulation of systemic phosphate homeostasis: is klotho an essential player? Am J Physiol Renal Physiol 2009;296:F470–F476.

8 Iotti S, Lodi R, Gottardi G, Zaniol P, Barbiroli B: Inorganic phosphate is transported into mitochondria in the absence of ATP biosynthesis: an in vivo 31P NMR study in the human skeletal muscle. Biochem Biophys Res Commun 1996;225:191–194.

9 Osuka S, Razzaque MS: Can features of phosphate toxicity appear in normophosphatemia? J Bone Miner Metab 2012;30:10–18.

10 Nagy AK, Kane DJ, Tran CM, Farley RA, Faller LD: Evidence calcium pump binds magnesium before inorganic phosphate. J Biol Chem 2005;280:7435–7443.

11 Pedemonte CH, Beauge L: Inhibition of (Na^+,K^+)-ATPase by magnesium ions and inorganic phosphate and release of these ligands in the cycles of ATP hydrolysis. Biochim Biophys Acta 1983;748:245–253.

12 Hansen O: The effect of sodium on inorganic phosphate- and *p*-nitrophenyl phosphate-facilitated ouabain binding to $(Na^+ + K^+)$-activated ATPase. Biochim Biophys Acta 1978; 511:10–22.

13 Reed JB, Smith SD, Forbes AB, Doxey DL: Inorganic phosphate, calcium and magnesium levels in the sera of Botswanan cattle receiving feed additives. Trop Anim Health Prod 1974;6:31–36.
14 Razzaque MS: Osteo-renal regulation of systemic phosphate metabolism. IUBMB Life 2011;63:240–247.
15 Berndt T, Kumar R: Novel mechanisms in the regulation of phosphorus homeostasis. Physiology (Bethesda) 2009;24:17–25.
16 Razzaque MS: The FGF23-klotho axis: endocrine regulation of phosphate homeostasis. Nat Rev Endocrinol 2009;5:611–619.
17 Marks J, Debnam ES, Unwin RJ: Phosphate homeostasis and the renal-gastrointestinal axis. Am J Physiol Renal Physiol 2010; 299:F285–F296.
18 Hattenhauer O, Traebert M, Murer H, Biber J: Regulation of small intestinal Na-Pi type IIb cotransporter by dietary phosphate intake. Am J Physiol Gastrointest Liver Physiol 1999;277:G756–G762.
19 Sabbagh Y, O'Brien SP, Song W, Boulanger JH, Stockmann A, Arbeeny C, Schiavi SC: Intestinal NPT2b plays a major role in phosphate absorption and homeostasis. J Am Soc Nephrol 2009;20:2348–2358.
20 Corut A, Senyigit A, Ugur SA, Altin S, Ozcelik U, Calisir H, Yildirim Z, Gocmen A, Tolun A: Mutations in SLC34A2 cause pulmonary alveolar microlithiasis and are possibly associated with testicular microlithiasis. Am J Hum Genet 2006;79:650–656.
21 Razzaque MS: Phosphate toxicity: new insights into an old problem. Clin Sci (Lond) 2011;120:91–97.
22 Miyamoto K, Ito M, Tatsumi S, Kuwahata M, Segawa H: New aspect of renal phosphate reabsorption: the type IIc sodium-dependent phosphate transporter. Am J Nephrol 2007; 27:503–515.
23 Tenenhouse HS: Phosphate transport: molecular basis, regulation and pathophysiology. J Steroid Biochem Mol Biol 2007;103:572–577.
24 Lorenz-Depiereux B, Benet-Pages A, Eckstein G, Tenenbaum-Rakover Y, Wagenstaller J, Tiosano D, Gershoni-Baruch R, Albers N, Lichtner P, Schnabel D, Hochberg Z, Strom TM: Hereditary hypophosphatemic rickets with hypercalciuria is caused by mutations in the sodium-phosphate cotransporter gene SLC34A3. Am J Hum Genet 2006;78:193–201.
25 Beck L, Karaplis AC, Amizuka N, Hewson AS, Ozawa H, Tenenhouse HS: Targeted inactivation of NPT2 in mice leads to severe renal phosphate wasting, hypercalciuria, and skeletal abnormalities. Proc Natl Acad Sci USA 1998;95:5372–5377.
26 Segawa H, Onitsuka A, Shiozawa K, Aranami F, Tatsumi S, Ito M, Kuwahata M, Miyamoto K: Roles of type IIc Na/Pi cotransporters in body pi homeostasis. J Am Soc Nephrol 2006; 17:355A.
27 Tenenhouse HS: Regulation of phosphorus homeostasis by the type IIa Na/phosphate cotransporter. Annu Rev Nutr 2005;25:197–214.
28 Biber J, Custer M, Magagnin S, Hayes G, Werner A, Lotscher M, Kaissling B, Murer H: Renal Na/Pi-cotransporters. Kidney Int 1996; 49:981–985.
29 Iharada M, Miyaji T, Fujimoto T, Hiasa M, Anzai N, Omote H, Moriyama Y: Type 1 sodium-dependent phosphate transporter (SLC17A1 protein) is a Cl(–)-dependent urate exporter. J Biol Chem 2010;285:26107–26113.
30 Suzuki A, Ammann P, Nishiwaki-Yasuda K, Sekiguchi S, Asano S, Nagao S, Kaneko R, Hirabayashi M, Oiso Y, Itoh M, Caverzasio J: Effects of transgenic Pit-1 overexpression on calcium phosphate and bone metabolism. J Bone Miner Metab 2010;28:139–148.
31 Festing MH, Speer MY, Yang HY, Giachelli CM: Generation of mouse conditional and null alleles of the type III sodium-dependent phosphate cotransporter Pit-1. Genesis 2009; 47:858–863.
32 Caldas Y, Giral H, Sorribas V, Levi M: Nuclear receptor LXR: a new partner for sodium-dependent phosphate cotransporters. Contrib Nephrol. Basel, Karger, 2013, vol 180, pp 64–73.
33 Kido S, Kaneko I, Tatsumi S, Segawa H, Miyamoto K: Vitamin D and type II sodium-dependent phosphate cotransporters. Contrib Nephrol. Basel, Karger, 2013, vol 180, pp 86–97.
34 Quarles LD: Endocrine functions of bone in mineral metabolism regulation. J Clin Invest 2008;118:3820–3828.
35 Baum M, Schiavi S, Dwarakanath V, Quigley R: Effect of fibroblast growth factor-23 on phosphate transport in proximal tubules. Kidney Int 2005;68:1148–1153.

36 Kanatani M, Sugimoto T, Kano J, Kanzawa M, Chihara K: Effect of high phosphate concentration on osteoclast differentiation as well as bone-resorbing activity. J Cell Physiol 2003;196:180–190.
37 Prie D, Torres PU, Friedlander G: Phosphate handling: new genes, new molecules. Horm Res Paediatr 2011;76(suppl 1):71–75.
38 Razzaque MS: The dualistic role of vitamin D in vascular calcifications. Kidney Int 2011;79: 708–714.
39 Razzaque MS: Does FGF23 toxicity influence the outcome of chronic kidney disease? Nephrol Dial Transplant 2009;24:4–7.
40 Drueke TB: Klotho, FGF23, and FGF receptors in chronic kidney disease: a yin-yang situation? Kidney Int 2010;78:1057–1060.
41 Razzaque MS, Lanske B: The emerging role of the fibroblast growth factor-23-klotho axis in renal regulation of phosphate homeostasis. J Endocrinol 2007;194:1–10.
42 Bastepe M, Juppner H: Inherited hypophosphatemic disorders in children and the evolving mechanisms of phosphate regulation. Rev Endocr Metab Disord 2008;9:171–180.
43 Benet-Pages A, Orlik P, Strom TM, Lorenz-Depiereux B: An FGF23 missense mutation causes familial tumoral calcinosis with hyperphosphatemia. Hum Mol Genet 2005;14: 385–390.
44 Goetz R, Beenken A, Ibrahimi OA, Kalinina J, Olsen SK, Eliseenkova AV, Xu C, Neubert TA, Zhang F, Linhardt RJ, Yu X, White KE, Inagaki T, Kliewer SA, Yamamoto M, Kurosu H, Ogawa Y, Kuro-o M, Lanske B, Razzaque MS, Mohammadi M: Molecular insights into the klotho-dependent, endocrine mode of action of fibroblast growth factor 19 subfamily members. Mol Cell Biol 2007;27:3417–3428.
45 Goetz R, Nakada Y, Hu MC, Kurosu H, Wang L, Nakatani T, Shi M, Eliseenkova AV, Razzaque MS, Moe OW, Kuro-o M, Mohammadi M: Isolated C-terminal tail of FGF23 alleviates hypophosphatemia by inhibiting FGF23-FGFR-klotho complex formation. Proc Natl Acad Sci USA 2010;107:407–412.
46 Goetz R, Ohnishi M, Kir S, Kurosu H, Wang L, Pastor J, Ma J, Gai W, Kuro-o M, Razzaque MS, Mohammadi M: Conversion of a paracrine fibroblast growth factor into an endocrine fibroblast growth factor. J Biol Chem 2012;287:29134–29146.
47 Goetz R, Ohnishi M, Ding X, Kurosu H, Wang L, Akiyoshi J, Ma J, Gai W, Sidis Y, Pitteloud N, Kuro OM, Razzaque MS, Mohammadi M: Klotho coreceptors inhibit signaling by paracrine fibroblast growth factor 8 subfamily ligands. Mol Cell Biol 2012;32:1944–1954.
48 Kuro-o M: Klotho as a regulator of fibroblast growth factor signaling and phosphate/calcium metabolism. Curr Opin Nephrol Hypertens 2006;15:437–441.
49 Medici D, Razzaque MS, Deluca S, Rector TL, Hou B, Kang K, Goetz R, Mohammadi M, Kuro OM, Olsen BR, Lanske B: FGF-23-klotho signaling stimulates proliferation and prevents vitamin D-induced apoptosis. J Cell Biol 2008;182:459–465.
50 Nabeshima Y: The discovery of alpha-klotho and FGF23 unveiled new insight into calcium and phosphate homeostasis. Cell Mol Life Sci 2008;65:3218–3230.
51 Razzaque MS: The role of klotho in energy metabolism. Nat Rev Endocrinol 2012;8:579–587.
52 Matsumura Y, Aizawa H, Shiraki-Iida T, Nagai R, Kuro-o M, Nabeshima Y: Identification of the human klotho gene and its two transcripts encoding membrane and secreted klotho protein. Biochem Biophys Res Commun 1998;242:626–630.
53 Maeda R, Imura A, Nabeshima Y: Complex regulation and diverse functions of a-klotho. Contrib Nephrol. Basel, Karger, 2013, vol 180, pp 25–46.
54 Nakatani T, Ohnishi M, Razzaque MS: Inactivation of klotho function induces hyperphosphatemia even in presence of high serum fibroblast growth factor 23 levels in a genetically engineered hypophosphatemic (HYP) mouse model. FASEB J 2009;23: 3702–3711.
55 Nakatani T, Sarraj B, Ohnishi M, Densmore MJ, Taguchi T, Goetz R, Mohammadi M, Lanske B, Razzaque MS: In vivo genetic evidence for klotho-dependent, fibroblast growth factor 23 (FGF23)-mediated regulation of systemic phosphate homeostasis. FASEB J 2009;23:433–441.
56 Ohnishi M, Kato S, Akiyoshi J, Atfi A, Razzaque MS: Dietary and genetic evidence for enhancing glucose metabolism and reducing obesity by inhibiting klotho functions. FASEB J 2011;25:2031–2039.

57 Ohnishi M, Nakatani T, Lanske B, Razzaque MS: In vivo genetic evidence for suppressing vascular and soft-tissue calcification through the reduction of serum phosphate levels, even in the presence of high serum calcium and 1,25-dihydroxyvitamin D levels. Circ Cardiovasc Genet 2009;2:583–590.
58 Ohnishi M, Nakatani T, Lanske B, Razzaque MS: Reversal of mineral ion homeostasis and soft-tissue calcification of klotho knockout mice by deletion of vitamin d 1alpha-hydroxylase. Kidney Int 2009;75:1166–1172.
59 Ohnishi M, Razzaque MS: Dietary and genetic evidence for phosphate toxicity accelerating mammalian aging. FASEB J 2010;24: 3562–3571.
60 Sitara D, Kim S, Razzaque MS, Bergwitz C, Taguchi T, Schuler C, Erben RG, Lanske B: Genetic evidence of serum phosphate-independent functions of FGF-23 on bone. PLoS Genet 2008;4:e1000154.
61 Ichikawa S, Imel EA, Kreiter ML, Yu X, Mackenzie DS, Sorenson AH, Goetz R, Mohammadi M, White KE, Econs MJ: A homozygous missense mutation in human klotho causes severe tumoral calcinosis. J Clin Invest 2007;117:2684–2691.
62 Bai X, Miao D, Li J, Goltzman D, Karaplis AC: Transgenic mice overexpressing human fibroblast growth factor 23(r176q) delineate a putative role for parathyroid hormone in renal phosphate wasting disorders. Endocrinology 2004;145:5269–5279.
63 Larsson T, Marsell R, Schipani E, Ohlsson C, Ljunggren O, Tenenhouse HS, Juppner H, Jonsson KB: Transgenic mice expressing fibroblast growth factor 23 under the control of the alpha1(I) collagen promoter exhibit growth retardation, osteomalacia, and disturbed phosphate homeostasis. Endocrinology 2004;145:3087–3094.
64 Shimada T, Urakawa I, Yamazaki Y, Hasegawa H, Hino R, Yoneya T, Takeuchi Y, Fujita T, Fukumoto S, Yamashita T: FGF-23 transgenic mice demonstrate hypophosphatemic rickets with reduced expression of sodium phosphate cotransporter type IIa. Biochem Biophys Res Commun 2004;314:409–414.
65 DeLuca S, Sitara D, Kang K, Marsell R, Jonsson K, Taguchi T, Erben RG, Razzaque MS, Lanske B: Amelioration of the premature aging-like features of FGF-23 knockout mice by genetically restoring the systemic actions of FGF-23. J Pathol 2008;216:345–355.
66 Bai X, Dinghong Q, Miao D, Goltzman D, Karaplis AC: Klotho ablation converts the biochemical and skeletal alterations in FGF23 (r176q) transgenic mice to a klotho-deficient phenotype. Am J Physiol Endocrinol Metab 2009;296:E79–E88.
67 Razzaque MS: Therapeutic potential of klotho-FGF23 fusion polypeptides: Wo2009095372. Expert Opin Ther Pat 2010.
68 Razzaque MS: FGF23, klotho and vitamin D interactions: what have we learned from in vivo mouse genetics studies? Adv Exp Med Biol 2012;728:84–91.
69 Strom TM, Francis F, Lorenz B, Boddrich A, Econs MJ, Lehrach H, Meitinger T: Pex gene deletions in Gy and Hyp mice provide mouse models for X-linked hypophosphatemia. Hum Mol Genet 1997;6:165–171.
70 Econs MJ, Friedman NE, Rowe PS, Speer MC, Francis F, Strom TM, Oudet C, Smith JA, Ninomiya JT, Lee BE, Bergen H: A PHEX gene mutation is responsible for adult-onset vitamin D-resistant hypophosphatemic osteomalacia: evidence that the disorder is not a distinct entity from X-linked hypophosphatemic rickets. J Clin Endocrinol Metab 1998;83:3459–3462.
71 Titan SM, Zatz R, Graciolli FG, dos Reis LM, Barros RT, Jorgetti V, Moyses RM: FGF-23 as a predictor of renal outcome in diabetic nephropathy. Clin J Am Soc Nephrol 2011;6: 241–247.
72 Slatopolsky E: The intact nephron hypothesis: the concept and its implications for phosphate management in CKD-related mineral and bone disorder. Kidney Int Suppl 2011;79(S121):S3–S8.
73 Martin KJ, Gonzalez EA: Prevention and control of phosphate retention/hyperphosphatemia in CKD-MBD: what is normal, when to start, and how to treat? Clin J Am Soc Nephrol 2011;6:440–446.
74 Hruska KA, Mathew S: The roles of the skeleton and phosphorus in the CKD mineral bone disorder. Adv Chron Kid Dis 2011;18: 98–104.
75 Fukagawa M, Kazama JJ: With or without the kidney: the role of FGF23 in CKD. Nephrol Dial Transplant 2005;20:1295–1298.
76 Imanishi Y, Inaba M, Nakatsuka K, Nagasue K, Okuno S, Yoshihara A, Miura M, Miyauchi A, Kobayashi K, Miki T, Shoji T, Ishimura E, Nishizawa Y: FGF-23 in patients with end-stage renal disease on hemodialysis. Kidney Int 2004;65:1943–1946.

77 Nishi H, Nii-Kono T, Nakanishi S, Yamazaki Y, Yamashita T, Fukumoto S, Ikeda K, Fujimori A, Fukagawa M: Intravenous calcitriol therapy increases serum concentrations of fibroblast growth factor-23 in dialysis patients with secondary hyperparathyroidism. Nephron Clin Pract 2005;101:c94–c99.

78 Liu S, Tang W, Zhou J, Stubbs JR, Luo Q, Pi M, Quarles LD: Fibroblast growth factor 23 is a counter-regulatory phosphaturic hormone for vitamin d. J Am Soc Nephrol 2006;17: 1305–1315.

79 Shigematsu T, Kazama JJ, Yamashita T, Fukumoto S, Hosoya T, Gejyo F, Fukagawa M: Possible involvement of circulating fibroblast growth factor 23 in the development of secondary hyperparathyroidism associated with renal insufficiency. Am J Kidney Dis 2004;44: 250–256.

80 Gutierrez O, Isakova T, Rhee E, Shah A, Holmes J, Collerone G, Juppner H, Wolf M: Fibroblast growth factor-23 mitigates hyperphosphatemia but accentuates calcitriol deficiency in chronic kidney disease. J Am Soc Nephrol 2005;16:2205–2215.

81 Li YC: Vitamin D in chronic kidney disease. Contrib Nephrol. Basel, Karger, 2013, vol 180, pp 98–109.

82 Nakanishi S, Fukagawa M: Functional disorder of the parathyroid gland and vascular calcification in hemodialysis patients. Clin Calcium 2005;15:119–123.

83 Koizumi M, Komaba H, Fukagawa M: Parathyroid function in chronic kidney disease: role of FGF23-klotho axis. Contrib Nephrol. Basel, Karger, 2013, vol 180, pp 110–123.

84 Fliser D, Kollerits B, Neyer U, Ankerst DP, Lhotta K, Lingenhel A, Ritz E, Kronenberg F, Kuen E, Konig P, Kraatz G, Mann JF, Muller GA, Kohler H, Riegler P: Fibroblast growth factor 23 (FGF23) predicts progression of chronic kidney disease: The Mild to Moderate Kidney Disease (MMKD) study. J Am Soc Nephrol 2007;18:2600–2608.

85 Hsu HJ, Wu MS: Fibroblast growth factor 23: a possible cause of left ventricular hypertrophy in hemodialysis patients. Am J Med Sci 2009;337:116–122.

86 Gutierrez OM, Januzzi JL, Isakova T, Laliberte K, Smith K, Collerone G, Sarwar A, Hoffmann U, Coglianese E, Christenson R, Wang TJ, deFilippi C, Wolf M: Fibroblast growth factor 23 and left ventricular hypertrophy in chronic kidney disease. Circulation 2009;119:2545–2552.

87 Takeshita K, Fujimori T, Kurotaki Y, Honjo H, Tsujikawa H, Yasui K, Lee JK, Kamiya K, Kitaichi K, Yamamoto K, Ito M, Kondo T, Iino S, Inden Y, Hirai M, Murohara T, Kodama I, Nabeshima Y: Sinoatrial node dysfunction and early unexpected death of mice with a defect of klotho gene expression. Circulation 2004;109:1776–1782.

88 Hu MC, Kuro-o M, Moe OW: Klotho and chronic kidney disease. Contrib Nephrol. Contrib Nephrol. Basel, Karger, 2013, vol 180, pp 47–63.

89 Shutto Y, Shimada M, Kitajima M, Yamabe H, Razzaque MS: Lack of awareness among future medical professionals about the risk of consuming hidden phosphate-containing processed food and drinks. PLoS One 2011; 6:e29105.

90 Raggi P, Bellasi A: Clinical assessment of vascular calcification. Adv Chronic Kidney Dis 2007;14:37–43.

91 Razzaque MS: Phosphate toxicity and vascular mineralization. Contrib Nephrol. Basel, Karger, 2013, vol 180, pp 74–85.

92 Michigami T: Extracellular phosphate as a signaling molecule. Contrib Nephrol. Basel, Karger, 2013, vol 180, pp 14–24.

93 Uchihashi K, Nakatani T, Goetz T, Mohammadi M, He X, Razzaque MS: FGF23-induced hypophosphatemia persists in HYP mice deficient in the WNT co-receptor LRP6. Contrib Nephrol. Basel, Karger, 2013, vol 180, pp 124–137.

94 Hartman M-L, Groppo F, Ohnishi M, Goodson JM, Hatice H, Tavares M, Yaskell T, Floros C, Behbehani K, Razzaque MS: Can salivary phosphate levels be an early biomarker to monitor the evolvement of obesity? Contrib Nephrol. Basel, Karger, 2013, vol 180, pp 138–148.

Mohammed S. Razzaque, MBBS, PhD
Department of Oral Medicine, Infection and Immunity
Harvard School of Dental Medicine, Research and Education Building
Room 304, 190 Longwood Avenue, Boston, MA 02115 (USA)
E-Mail mrazzaque@hms.harvard.edu, debpur@gmail.com

Razzaque MS (ed): Phosphate and Vitamin D in Chronic Kidney Disease.
Contrib Nephrol. Basel, Karger, 2013, vol 180, pp 14–24 (DOI: 10.1159/000346776)

Extracellular Phosphate as a Signaling Molecule

Toshimi Michigami

Department of Bone and Mineral Research, Osaka Medical Center and Research Institute for Maternal and Child Health, Izumi, Osaka, Japan

Abstract

Phosphorus is involved in various biological processes including membrane integrity, maintenance and inheritance of genetic materials, energy metabolism, intracellular signaling and skeletal mineralization. In addition, accumulating evidences have indicated that alteration in the levels of extracellular inorganic phosphate (Pi) itself triggers signaling to regulate gene expression and cellular functions in some cell types. In bone cells such as osteoblasts and chondrocytes, extracellular Pi modulates cell proliferation, differentiation, mineralization and apoptosis. In extraskeletal tissues, extracellular Pi also exerts various effects. For example, increased extracellular Pi results in the calcification associated with the upregulation of osteoblast marker genes in vascular smooth muscle cells. As to the mechanistic aspects, it is suggested that an increase in extracellular Pi triggers signal transduction via the PiT1 type III sodium/phosphate (Na^+/Pi) cotransporter and ERK1/2 pathway. Unicellular organisms such as bacteria and yeast sense the environmental Pi with a protein complex located in the plasma membrane, which regulates the expression of multiple genes involved in Pi uptake and metabolism to adapt to its availability. In mammals that are multicellular organisms, Pi availability should be sensed both at a cellular level to regulate the function of each cell and as a whole body to maintain the Pi homeostasis of the extracellular fluid. Although the responsiveness to the increased extracellular Pi suggests the existence of Pi-sensing mechanism in mammalian cells as well, it is unknown whether the sensing of Pi availability at a cellular level and that at a whole-body level share the same pathway or not. This chapter will review the findings regarding the regulation of various cellular functions by extracellular Pi, and also discuss the current concept on the mechanism for Pi-sensing.

Phosphorus is critical in various biological processes including membrane integrity, synthesis of deoxyribonucleic acid (DNA) and ribonucleic acid (RNA), energy metabolism and intracellular signaling. In vertebrates, phosphorus is also essential in skeletal mineralization. In human adults, approximately 85% of the total phosphorus exists in bone as complex with calcium in hydroxyapatite crystals deposited onto the collagen matrix. Fifteen percent of phosphorus is present in soft tissues with only 0.1% in the extracellular fluids [1]. Phosphorus in serum is present mainly as inorganic phosphate (Pi) in the form of free ions HPO_4^{2-} and $H_2PO_4^-$, and the former is dominant in physiological pH. The majority of the intracellular phosphate ion exists as bound form or inorganic phosphate esters, phospholipids, or phosphorylated intermediate molecules involved in various biochemical processes. The concentration of Pi in the extracellular fluid and cytosol is about 10^{-4} and 2×10^{-4} M, respectively. The intracellular concentration of Pi is maintained by membrane transporters, which function dependently on sodium (Na^+) gradient across the plasma membrane [1].

In mammals, Na^+-coupled Pi cotransporters have been classified into three families, based on the amino acid sequence and structural similarity [2]. All cotransporters are multispanning membrane proteins. The type I Na^+/Pi cotransporter belongs to the solute carrier family 17 (SLC17), and is known as NPT1 in humans. In addition to the Pi transport activity, type I cotransporters transport organic anions, and their physiological roles are not fully established. Type II Na^+/Pi cotransporters are assigned to SLC34, and include NPT2a, NPT2b and NPT2c. Npt2a and Npt2c were shown to be predominantly expressed at the brush border membrane of proximal tubular epithelial cells, and responsible for the renal reabsorption of Pi. On the other hand, Npt2b is expressed in various tissues including small intestine, and has been revealed to be responsible for intestinal apical Na^+/Pi cotransport. Type III Na^+/Pi cotransporters are classified as SLC20 family, which includes PiT1 (SLC20A1) and PiT2 (SLC20A2). These proteins are widely expressed, and were originally identified as retroviral receptors Glvr-1 and Ram-1, respectively [2].

In addition to the profound roles of phosphorus in various biological processes, it has been implicated that extracellular Pi itself functions as a signaling molecule that regulates gene expression and modulates cell behavior. This chapter will review the recent findings on the regulation of cellular function by extracellular Pi. In addition, current concept on the sensing mechanism of environmental Pi will be discussed.

Roles of Extracellular Pi in Skeleton

In mammals, renal reabsorption and intestinal absorption of Pi are central to control Pi balance [1]. Prolonged deficiency of Pi due to the abnormal absorption or reabsorption results in the poorly mineralized bone characteristic of rickets or osteomalacia, indicating the critical role of phosphorus in skeletal mineralization. Recently, it has been revealed that extracellular Pi regulates the various cellular functions in skeleton other than mineralization.

Alkaline phosphatase is the enzyme that hydrolyses pyrophosphates and the protein-bound form of phosphate to produce Pi and increase its concentration in the extracellular microenvironment. During the early phase of osteoblast differentiation, the expression of the gene for alkaline phosphatase gradually increases. Using a murine osteoblastic cell line MC3T3-E1 cells, Beck et al. [3] found that a decrease in extracellular Pi production due to the inhibition of alkaline phosphatase activity caused a decreased expression of osteopontin, which is involved in the regulation of mineralization. Then, they further demonstrated that the increased Pi levels in the culture medium strongly induced the expression of osteopontin, and firstly indicated that extracellular Pi itself functions as a specific signal that regulates the gene expression in osteoblastic cells [4]. The authors also performed a combined proteome and microarray investigation in MC3T3-E1 cells to identify the molecules comprehensively whose expression levels were altered in response to treatment with the higher concentration of extracellular Pi [5]. Among the molecules whose amount was increased in response to Pi, some were found to be involved in the regulation of cell cycle, proliferation and DNA synthesis [5]. Thus, the extracellular Pi in microenvironment seems to regulate both proliferation and differentiation of osteoblasts.

Extracellular Pi plays a role in regulating chondrogenesis as well. Chondrogenesis is a multistep process involving the mesenchymal condensation of undifferentiated cells, the proliferation of chondrocytes, and differentiation into hypertrophic chondrocytes, followed by mineralization. As chondrocytes mature, the concentration of Pi increases in the extracellular milieu. The progressive accumulation of Pi ion starts from the proliferative zone and peaks in the hypertrophic zone [6].

Terminally differentiated chondrocytes in growth plate undergo mineralization associated with apoptosis. Several studies have indicated that high level of Pi in the microenvironment of mature chondrocytes induces apoptosis [7–9]. Sabbagh et al. [7] demonstrated that the decreased level of extracellular Pi impaired the apoptosis of hypertrophic chondrocytes and lead to rickets in the hypophosphatemic *Hyp* mouse, a murine model for human X-linked hypophosphatemia. Their data implicate that extracellular Pi plays a critical role for the regulated apoptosis of hypertrophic chondrocytes via the activation of the caspase-9-mediated mitochondrial pathway. Mansfield et al. [8] linked the apoptosis of mature chondrocytes directly to the elevation in environmental Pi concentration and the concomitant rise in intracellular Pi levels, and showed that the effect of Pi on chondrocyte apoptosis was calcium dependent. More recently, Sugita et al. [9] reported that Pi-regulated apoptosis of hypertrophic chondrocytes involved the intracellular adenosine 5′-triphosphate synthesis mediated by type III Na^+/Pi cotransporter Pit1.

Wang et al. [10] examined the expression of Na^+/Pi cotransporters in a chondrocytic cell line CFK2 cells, and detected the expression of type III cotransporters Pit1 and Pit2 [7]. Transient increase in the expression of these genes was observed at 1 week of differentiation, which preceded maximum expression of markers of chondrocyte differentiation such as type II collagen [10]. Their data suggest that Pi may have some role in early chondrocytes as well as mature chondrocytes. We investigated the role of Pi in early chondrocytes at the proliferating stage and found that the treatment with higher concentration of extracellular Pi induced the upregulation of cyclin D1 within 24 h [11]. Cyclin D1 is a positive regulator of progression through the G1 phase of the cell cycle, and is induced to express by many mitogenic stimuli. Consistent with the upregulation of cyclin D1 expression, the extracellular Pi facilitated the proliferation in early chondrocytes [11].

The effects of extracellular Pi on the maturation of chondrocytes were also demonstrated. ATDC5 is a murine mesenchymal cell line and used as a model for chondrogenesis. In differentiated ATDC5 cells, Pi accelerated the expression of type X collagen, a marker for hypertrophic chondrocytes [12]. The expression of matrix Gla protein, which is related to mineralization, was also increased in response to high extracellular Pi [13]. Thus, the signals of extracellular Pi are involved in various processes in chondrogenesis including cell proliferation, maturation and apoptosis, and the effects might differ among the stages of the chondrocytes.

Effects of Extracellular Pi in Extraskeletal Tissues

Altered levels in extracellular Pi influence the gene expression and cell function in extraskeletal tissues as well. Kido et al. [14] reported that feeding mice with a low-phosphorus diet resulted in the increased level of *Npt2a* Na^+/Pi cotransporter in kidney, and identified a DNA sequence responsible for Pi in the *Npt2a* gene. The Pi-response element of the *Npt2a* gene was found to consist of the motif related to the E-box, 5′-CACGTG-3′. They cloned a transcription factor that bound to the Pi-response element by yeast one-hybrid system, and identified the mouse transcription factor muE3 (TFE3).

In vascular smooth muscle cells, a higher concentration of extracellular Pi induced the expression of osteoblast-specific genes such as *Runx2* and *osteocalcin*, and caused the phenotypical change that predisposed to calcification [15, 16]. This altered gene expression by increased extracellular Pi was suggested to be mediated by Pit1 [16]. Shuto et al. [17] investigated the effects of high dietary phosphorus in healthy human adults, and found that high dietary phosphorus load resulted in the increased the serum Pi levels at 2 h after the meal and acutely impaired the endothelial function. These results suggest the involvement of high extracellular Pi in the pathogenesis of cardiovascular events associated with chronic kidney disease.

Extracellular Pi directly exerts its effects on the parathyroid gland as well. It was reported that the elevated level of Pi in media increased the parathyroid hormone (PTH) secretion from fresh rat parathyroid grand in vitro [18]. Thus, a great variety of cells are able to respond to the environmental Pi.

How Is the Pi-Induced Signal Transduced?

As described above, extracellular Pi itself modulates the functions in various cells. Then, how is the extracellular Pi-triggered signal transduced into the cells? It was demonstrated that the increased extracellular Pi induced the activation of ERK1/2 pathway, but not that of the other pathways such as p38MAPK pathway or JNK pathway, in several cells including MC3T3-E1 cells and ATDC5 cells [11, 13, 19]. In these cells, inhibition of the ERK1/2 pathway abolished the regulation of the genes by extracellular Pi, indicating its importance in the transduction of the signals triggered by extracellular Pi.

Another important question is whether Pi has to enter the cell to exert its biological effects or not. It was reported that treatment with phosphonoformic acid (PFA), a well-documented competitive inhibitor of SLC34 family of Na^+/Pi cotransporters, prevented the calcification of vascular smooth muscle cells expressing Pit1 and Pit2 by inhibiting calcium-phosphate deposition [20]. Since PFA hardly inhibits the Pi transport mediated by SLC20 (PiT) family cotransporters [2], its effect on vascular smooth muscle calcification might be independent of Pi uptake. On the other hand, Li et al. [16] showed that knockdown of the expression of *Pit1* impeded the Pi-induced calcification of vascular smooth muscle cells. In our study, knockdown of *Pit1* in ATDC5 cells abrogated the Pi-induced activation of the Raf/MEK/ERK pathway, suggesting involvement of Pit1 in the responsiveness to extracellular Pi [11]. These results indicate the involvement of Pit1 in transduction of the signals triggered by altered levels of extracellular Pi in some cell types. However, it still remains unclear whether the entry into the cell is required or not for extracellular Pi to trigger signal transduction. As described above, transporters of Pit family were initially identified as receptors for retroviruses, and recent studies have demonstrated some roles of Pit1 that are independent of its role in Pi transport [21]. Further studies are needed to uncover the precise mechanism by which Pit1 mediates the effects of altered extracellular Pi.

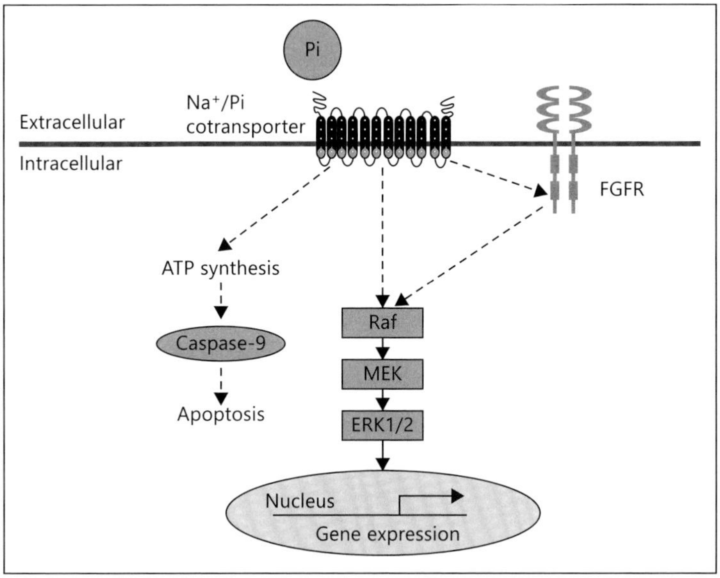

Fig. 1. Transduction of the signal triggered by extracellular Pi. In several cell types such as osteoblastic MC3T3-E1 cells and chondrocytic ATDC5 cells, increased extracellular Pi induces the activation of Raf/MEK/ERK1/2 pathway and regulates gene expression, a process that is mediated by the Na^+/Pi cotransporter. In human embryonic kidney cell line HEK293 cells, FGFR is involved in the Pi-induced activation of the ERK1/2 pathway. In hypertrophic chondrocytes, increased extracellular Pi induces apoptosis via intracellular ATP synthesis and caspase-9-mediated mitochondrial pathway.

In a human embryonic kidney cell line HEK293, we found that an increase in extracellular Pi induced the phosphorylation fibroblast growth factor receptor substrate 2α (FRS2α) as well as its downstream molecule ERK1/2. Knockdown of *fibroblast growth factor receptor 1* (*FGFR1*) expression as well as that of *PiT1* diminished the phosphorylation of both FRS2α and ERK1/2 induced by extracellular Pi. Moreover, overexpression of FGFR1 rescued the decrease in Pi-induced phosphorylation of ERK1/2 in the cells where the expression of *PiT1* was knocked down [22]. These results suggest that increased extracellular Pi triggers signal transduction via PiT1 and FGFR in HEK293 cells, although the mechanism that mediates the signaling from PiT1 to FGFR1 remains to be elucidated (fig. 1).

Data on the transcription factors that might be involved in the gene regulation by extracellular Pi in mammals are limited. As described above, TFE3 was identified to bind to the Pi-response element in mouse *Npt2a* gene [14]. Early growth response-1, a downstream transcription factor of ERK1/2, was upregulated in response to the increased extracellular Pi in HEK293 cells [22]. In MC3T3-E1 cells, Beck et al. [23] found the transcription factor Nrf2 to be regulated by extracellular

Pi. Moreover, by a combined proteome and transcriptome analyses in MC3T3-E1, they identified that Fra-1, a member of the AP-1 family of transcription factors, was not regulated at the mRNA level, but its protein level was increased in response to extracellular Pi [5]. It was reported that the regulation of matrix Gla protein expression by extracellular Pi in differentiated ATDC5 cells involved the facilitation of the DNA binding of Fra-1 to specific sequences in the matrix Gla protein promoter [13]. It is unknown whether a transcription factor exists or not that functions specifically to mediate the signals of extracellular Pi in mammals.

Sensing of Environmental Pi in Bacteria and Yeast

Responsiveness to the alteration in the levels of extracellular Pi suggests the existence of the mechanism by which the cells sense it. In unicellular cells such as bacteria and yeast, the systems for sensing the environmental Pi have been extensively studied.

In bacteria, environmental Pi is sensed by ATP-binding cassette (ABC)-type Pi specific transporter (Pst) and a protein called PhoU. The Pst transporter is composed of four subunits, PstS, PstA, PstB and PstC. Pi binds to Pst, and activates a two-component system (TCS) composed of sensory histidine kinase (HK) PhoR, which is an integral membrane protein, and its partner DNA-binding response regulator (RR) PhoU, which is a transcription factor. Under conditions of Pi limitation, phosphorylation of the histidine residues in PhoR is increased, and then the phosphorylated PhoR in turn activates PhoB by phosphorylation on its aspartic residues. The phosphorylated PhoB increases the transcription of multiple genes in the Pho regulon by binding to the specific DNA sequences called PHO boxes, which allows the bacteria to adapt the reduced Pi concentration. If the Pi concentration in the environment is adequate, PhoR is dephophorylated, and PhoB remains unphosphorylated and is incapable of inducing the transcription of the genes in the Pho regulon. Estimates for the number of Pi-regulated genes vary widely, but proteomic data revealed that amounts of nearly 400 proteins altered in response to the environmental Pi level [24, 25].

Similar to the case of bacteria, environmental Pi controls gene expression through a specialized transcription factor, named Pho4, in yeast. The activity and subcellular localization of Pho4 is regulated by the cyclin-dependent kinase complex Pho80-Pho85. Reduced level of environmental Pi results in the inactivation of the Pho80-Pho85 by a cyclin-dependent kinase inhibitor Pho81. Consequently, Pho4, which remains unphosphorylated and active, translocates into the nucleus to induce the transcription of the target genes belonging to the yeast Pho regulon. One of them encodes Pho84 protein, which functions as a high-affinity

Pi transporter. Under the Pi-starved condition, the expression of Pho84 is strongly induced. Pho84 belongs to the major facilitator superfamily and catalyzes a bidirectional proton (H^+)-coupled Pi uptake, where the direction of transport is determined by the directionality of the driving force rather than by the orientation of the protein. When adequate level of Pi exists in the environment, Pho84 is rapidly internalized from the plasma membrane and degraded, a process that involves the activation of protein kinase A pathway. The degradation of Pho84 leads to phosphorylation of Pho4. The phosphorylated Pho4 is exported from the nucleus, and the expression of PHO genes is turned off [24, 26].

Pi-Sensing to Maintain Homeostasis in Mammals

Although various mammalian cells respond to the altered levels of extracellular Pi in vitro, little is known about how mammals sense the level of Pi. In multicellular organisms, Pi levels should be sensed at a whole-body level to maintain the homeostasis of extracellular Pi, as well as at a cellular level to regulate the effects of Pi on cell function. Pi homeostasis in mammals is controlled by a variety of hormones that regulate the renal reabsorption and intestinal absorption, including FGF23, PTH, and 1,25-dihydroxyvitamin D [27]. The availability of Pi at a whole-body level should be sensed to control the production of these hormones.

FGF23 is produced by osteocytes and osteoblasts in bone, and plays the central role in Pi homeostasis by reducing the Pi reabsorption and suppressing the synthesis of 1,25-dihydroxyvitamin D in kidney. Reflecting the endocrine function of FGF23, its circulating levels are controlled by various systemic factors, of which 1,25-dihydroxyvitamin D has the best defined regulatory role [28]. Although the extracellular Pi itself might regulate the production of FGF23, an increase in extracellular Pi did not directly stimulate FGF23 expression in osteoblasts [28]. Because of the lack of suitable in vitro cell culture models, it remains unclear whether Pi directly regulates the expression and/or secretion of FGF23. In addition to FGF23, osteocytes express dentin matrix protein 1 (DMP1) and phosphate regulating gene with homologies to endopeptidases on the X-chromosome (PHEX). These three molecules are responsible for inherited hypophosphatemic rickets/osteomalacia, and the loss of function in DMP1 or PHEX results in the increased level of FGF23, suggesting the intimate functional relationship among them [27]. Further studies on the regulatory mechanisms of these molecules might clarify the mechanism for sensing the Pi levels.

Martin et al. [29] investigated the response of serum PTH to changes dietary Pi using rats, and found that duodenal infusion of Pi increased serum PTH within minutes. Intravenous administration of Pi also rapidly increased serum PTH, suggesting the existence of Pi-sensing mechanism in parathyroid gland [29]. On the other hand, Berndt et al. [30] suggested the sensing of Pi in the intestine. They administered Pi into the duodenum of intact or parathyroidectomized rats, and examined the renal Pi excretion. The duodenal Pi loading resulted in the increased renal Pi excretion without alteration in the levels of PTH and FGF23, and the parathyroidectomy did not influence the response to Pi loading [30]. Their results suggest that the intestine can sense the increase in the luminal Pi concentration, although how the signal leads to increase in renal Pi excretion remains to be elucidated.

Conclusion

Phosphorus is an essential element for all living organisms. Unicellular organisms such as bacteria and yeast have been revealed to adapt the concentration of environmental Pi by sensing it through a system involving a protein complex in plasma membrane and a specialized transcription factor. Various mammalian cells also respond to the alteration of the extracellular Pi levels, and the signal triggered by Pi was found to be transduced through Na^+/Pi cotransporter and ERK1/2 pathway. Although the responsiveness to extracellular Pi suggests the existence of mammalian Pi-sensing system, Pi availability should be sensed both at a cellular level and at a whole-body level in mammals. Further studies will be required to clarify the underlying mechanisms of phosphate metabolism [31].

References

1 Favus MJ, Bushinskey DA, Lemann J Jr: Regulation of calcium, magnesium, and phosphate metabolism; in Favus MJ (ed): Primers on the Metabolic Bone Diseases and Disorders of the Mineral Metabolism, ed 6. Washington, American Society of Bone and Mineral Research, 2006, pp 76–83.

2 Virkki LV, Biber J, Murer H, Forster IC: Phosphate transporters: a tale of two solute carrier families. Am J Physiol Renal Physiol 2007;293:F643–F654.

3 Beck GR Jr, Sullivan EC, Moran E, Zerler B: Relationship between alkaline phosphtase levels, osteopontin expression, and mineralization in differentiating MC3T3-E1 osteoblasts. J Cell Biochem 1998;68:269–280.

4 Beck GR Jr, Zerler B, Moran E: Phosphate is a specific signal for induction of osteopontin gene expression. Proc Natl Acad Sci USA 2000;97:8352–8357.

5 Conrads KA, Yi M, Simpson KA, Lucas DA, Camalier CE, Yu L, Veenstra TD, Stephens RM, Conrads TP, Beck GR Jr: A combined proteome and microarray investigation of inorganic phosphate-induced pre-osteoblast cells. Mol Cell Proteomics 2005;4:1284–1296.

6 Mwale F, Tchetina E, Wu CW, Poole AR: The assembly and remodeling of the extracellular matrix in the growth plate in relationship to mineral deposition and cellular hypertrophy: an in situ study of collagens II and IX and proteoglycan. J Bone Miner Res 2002; 17:275–283.
7 Sabbagh Y, Carpenter TO, Demay MB: Hypophosphatemia leads to rickets by impairing caspase-mediated apoptosis of hypertrophic chondrocytes. Proc Natl Acad Sci USA 2005; 102:9637–9642.
8 Mansfield K, Pucci B, Adams CS, Shapiro IM: Induction of apoptosis I skeletal tissues: phosphate-mediated chick chondrocyte apoptosis is calcium-dependent. Calcif Tissue Int 2003;73:161–172.
9 Sugita A, Kawai S, Hayashibara T, Amano A, Ooshima T, Michigami T, Yoshikawa H, Yoneda T: Cellular ATP synthesis mediated by type III sodium-dependent phosphate transporter Pit-1 is critical to chondrogenesis. J Biol Chem 2011;286:3094–3103.
10 Wang D, Canaff L, Davidson D, Corluka A, Liu H, Hendy GN, Henderson JE: Alterations in the sensing and transport of phosphate and calcium by differentiating chondrocytes. J Biol Chem 2001;276:33995–34005.
11 Kimata M, Michigami T, Tachikawa K, Okada T, Koshimizu T, Yamazaki M, Kogo M, Ozono K: Signaling of extracellular inorganic phosphate upregulates cyclin D1 expression in proliferating chondrocytes via the Na^+/Pi cotransporter Pit-1 and Raf/MEK/ERK pathway. Bone 2010;47:938–947.
12 Magne D, Bluteau G, Faucheux C, Palmer G, Vignes-Colombeix C, Pilet P, Rouillon T, Caverzasio J, Weiss P, Daculsi G, Guicheux J: Phosphate is a specific signal for ATDC5 chondrocyte maturation and apoptosis-associated mineralization: possible implication of apoptosis in the regulation of endochondral ossification. J Bone Miner Res 2003;18:1430–1442.
13 Julien M, Magne D, Masson M, Rolli-Derkinderen M, Chassande O, Cario-Toumaniantz C, Cherel Y, Weiss P, Guicheux J: Phosphate stimulates matrix Gla protein expression in chondrocytes through the extracellular signal regulated kinase signaling pathway. Endocrinology 2007;148:530–537.
14 Kido S, Miyamoto K, Mizobuchi H, Taketani Y, Ohkido I, Ogawa N, Kaneko Y, Harashima S, Takeda E: Identification of regulatory sequences and binding proteins in the type II sodium/phosphate cotransporter NPT2 gene responsive to dietary phosphate. J Biol Chem 1999;274:28256–28263.
15 Jono S, McKee MD, Murry CE, Shioi A, Nishizawa Y, Mori K, Morii H, Giachelli CM: Phosphate regulation of vascular smooth muscle cell calcification. Circ Res 2000; 87:E10–E17.
16 Li X, Yang HY, Giachelli CM: Role of the sodium-dependent phosphate cotransporter, Pit-1, in vascular smooth muscle cell calcification. Circ Res 2006;98:905–912.
17 Shuto E, Taketani Y, Tanaka R, Harada N, Isshiki M, Sato M, Nashiki K, Amo K, Yamamoto H, Higashi Y, Nakaya Y, Takeda E: Dietary phosphorus acutely impairs endothelial function. J Am Soc Nephrol 2009;20:1504–1512.
18 Almaden Y, Canalejo A, Hernandez A, Ballesteros E, Garcia-Navarro S, Torres A, Rodriguez M: Direct effect of phosphorus on PTH secretion from whole rat parathyroid glands in vitro. J Bone Miner Res 1996;11:970–976.
19 Beck GR Jr, Knecht N: Osteopontin regulation by inorganic phosphate is ERK1/2-, protein kinase C-, and proteasome-dependent. J Biol Chem 2003;278:41921–41929.
20 Villa-Bellosta R, Sorribas V: Phosphonoformic acid prevents vascular smooth muscle cell calcification by inhibiting calcium-phosphate deposition. Arterioscler Thromb Vasc Biol 2009;29:761–766.
21 Beck L, Leroy C, Salaün C, Margall-Ducos G, Desdouets C, Friedlander G: Identification of a novel function of PiT1 critical for cell proliferation and independent of its phosphate transport activity. J Biol Chem 2009;45: 31363–31374.
22 Yamazaki M, Ozono K, Okada T, Tachikawa K, Kondou H, Ohata Y, Michigami T: Signal transduction triggered by an increase in extracellular inorganic phosphate involves fibroblast growth factor receptors and interacts with the signaling of fibroblast growth factor 23. J Cell Biochem 2010;111:1210–1221.
23 Beck GR Jr, Moran E, Knecht N: Inorganic phosphate regulates multiple genes during osteoblast differentiation, including Nrf2. Exp Cell Res 2003;288:288–300.
24 Bergwitz C, Jüppner H: Phosphate sensing. Adv Chron Kid Dis 2011;18:132–144.

25 Hsieh YJ, Wanner BL: Global regulation by the seven-component Pi signaling system. Curr Opin Microbiol 2010;13:198–203.
26 Mouillon JM, Persson BL: New aspects on phosphate sensing and signaling in *Saccharomyces cerevisiae*. FEMS Yeast Res 2006;6: 171–176.
27 Quarles LD: Endocrine functions of bone in mineral metabolism regulation. J Clin Invest 2008;118:3820–3828.
28 Liu S, Tang W, Zhou J, Stubbs JR, Luo Q, Pi M, Quarles LD: Fibroblast growth factor 23 is a counter-regulatory phosphaturic hormone for vitamin D. J Am Soc Nephrol 2006;17: 1305–1315.
29 Martin DR, Ritter CS, Slatopolsky E, Brown AJ: Acute regulation of parathyroid hormone by dietary phosphate. Am J Physiol Endocrinol Metab 2005;289:E729–E734.
30 Berndt TJ, Schiavi S, Kumar R: 'Phosphatonins' and the regulation of phosphorus homeostasis. Am J Physiol Endocrinol Metab 2005;289:F1170–F1182.
31 Ohnishi M, Razzaque MS: Osteo-renal cross-talk and phosphate metabolism by fgf23-klotho system. Contrib Nephrol. Basel, Karger, 2013, vol 180, pp 1–13.

Toshimi Michigami, MD, PhD
Department of Bone and Mineral Research
Osaka Medical Center and Research Institute for Maternal and Child Health
840 Murodo-cho, Izumi, Osaka 594-1101 (Japan)
E-Mail michigami@mch.pref.osaka.jp

Razzaque MS (ed): Phosphate and Vitamin D in Chronic Kidney Disease.
Contrib Nephrol. Basel, Karger, 2013, vol 180, pp 25–46 (DOI: 10.1159/000346777)

Complex Regulation and Diverse Functions of Alpha-Klotho

Ryota Maeda · Akihiro Imura · Yo-ichi Nabeshima

Laboratory of Molecular Life Science, Foundation for Biomedical Research and Innovation, Kobe, Japan

Abstract

It has been understood that the α-klotho gene, first identified as an aging-related gene, is actually necessary for regulating mineral homeostasis in vertebrates. All vertebrates, including humans, actively maintain calcium and phosphate ions in bones, circulating blood, and cerebrospinal fluid. Therefore, disruptions in homeostasis cause osteoporosis, ectopic calcification, and epilepsy. Vitamin D and parathyroid hormone are well-known hormones for maintaining calcium and phosphate balance. Synthesis of vitamin D, secretion of parathyroid hormone, and absorption of calcium and phosphate ions are all finely controlled by α-Klotho-dependent machineries. However, the precise molecular mechanisms and functions of α-Klotho are still unclear. In this article, we present an overview of recent progress in α-klotho research, with a main focus on molecular aspects.

In whole body fluids, concentrations of mineral ions, e.g. sodium ions, potassium ions, chloride ions, protons, hydrogen carbonate ions, heavy metal ions, calcium ions and phosphate ions, are tightly regulated within a certain ranges. If the concentrations of these ions decrease or increase in body fluids, several disorders could occur including hypertension, infarction, acidosis and calcification. Osteoporosis, skin atrophy, or arterial sclerosis is often observed especially in aged humans and is caused by their decreased ability to maintain calcium ion and phosphate ion homeostasis. Concentrations of calcium and phosphate ions are partially controlled via hormones, vitamin D and parathyroid hormone (PTH). Historically and clinically, it has been well known that these hormones are necessary for maintaining a calcium and phosphate ion balance within physiological concentrations. Furthermore, recent work has clarified that α-klotho, which was discovered as an aging-related gene, is also necessary for calcium and phosphate homeostasis.

In the aging process, irreversible damage to proteins and genomes accumulate in the whole body due to oxidative stress, ultraviolet damage, and chemicals. Consequently, any kind of homeostasis would not be able to be finely maintained. It can be said that the capability of maintaining homeostasis is directly connected to the aging process [1–7]. Therefore, α-klotho-deficient mice can be used as an aging model in mice, especially in mineral homeostatic disorders [8, 9]. Disorders such as osteoporosis, skin atrophy, and arterial sclerosis are all seen in α-klotho-deficient mice. Recent work showed that excessive vitamin D synthesis causes these aging-related phenotypes. Therefore, α-klotho can be considered more precisely as a gene which regulates mineral homeostasis [10, 11].

Evolution

A freshwater fish of teleosts, zebrafish (*Danio rerio*), has an α-klotho homologous gene [12]. The transcript is exclusively expressed in the gill, which is the origin of the parathyroid gland and where mineral ions are actively exchanged [13]. It has been reported that a putative homologous gene of mammalian α-klotho genes is expressed in the fruit fly (*Drosophila*) and nematode (*Caenorhabditis elegans*) [14, 15]. However, proposed α-klotho homologue genes are distinguished from mammalian α-klotho by a few points: the two genes (1) code cytosolic protein without signal peptides or transmembrane domains, (2) have only a KL1-like domain, and (3) conserves two catalytic glutamate residues. Therefore, it is not plausible that α-klotho homologous genes proposed in the fruit fly and nematode are indeed functional homologues of mammalian α-klotho genes. In these points, the two proposed genes are rather similar to the mammalian Klotho-related protein (KLrP) gene.

Molecular Aspects

The klotho gene, now renamed as the α-klotho gene (see below), was first described in 1997. The phenotype of mice deficient in the gene resembled that of aged humans [16]. The reported original mice were obtained as an α-klotho gene hypomorph strain, in which expression levels of the transcript were reduced due to an insertion mutation upstream of the gene. Then, α-klotho complete null mice were generated in 2003 [17]. The overall phenotype of α-klotho knockout

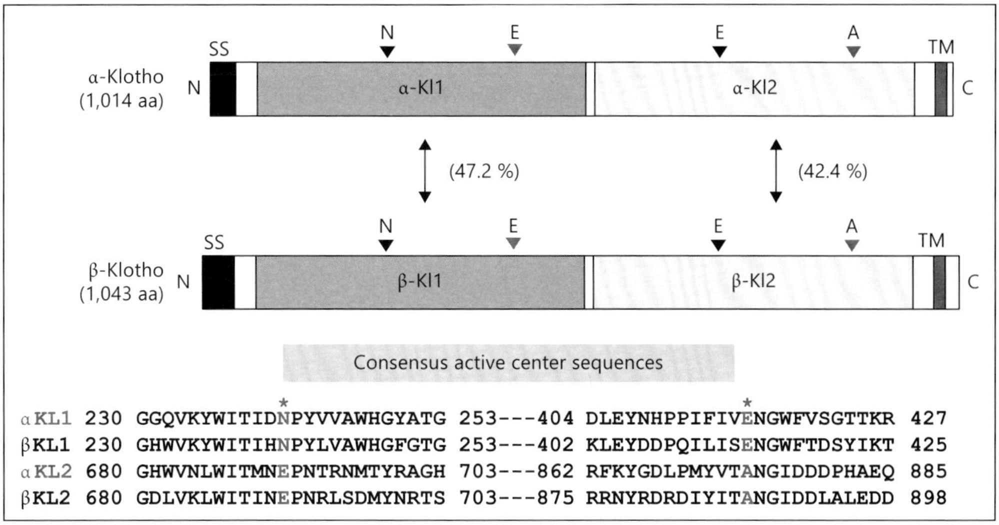

Fig. 1. Schematic diagrams of structural similarities of α-klotho and β-klotho. The acid/base and nucleophilic amino acids are indicated.

mice was similar to that of α-klotho hypomorph mice except that in α-klotho hypomorph mice, the protein expression of α-Klotho will be induced under some circumstances [18].

The transcript of the α-klotho gene encodes a single-pass type-I transmembrane protein with two external glycosidase-like domains. The translated protein is modified, most likely by N-glycosylation, and is observed as at least two molecular masses between 120 and 135 kDa [19]. A short splicing variant has been detected, but any translated proteins of the short transcript have not been observed at this time [19–21].

It is quite an interesting feature that two glutamate residues, which are indispensable for glycosidase activity, are not conserved in the two glycosidase-like domains of α-klotho [22, 23]. In the N-terminal KL1 domain, the acid/base glutamate is changed to asparagine and, in the C-terminal KL2 domain, nucleophilic glutamate is replaced by alanine or serine. In 2004, another high homologous gene of α-klotho was identified and named the β-klotho gene, which also encodes a single-pass type-I transmembrane protein with two external glycosidase-like domains. Catalytic glutamate residues are replaced in the same way as above [24, 25]. Therefore, we have referred to the original klotho gene as the α-klotho gene according to natural nomenclature (fig. 1).

Structures

α-Klotho was classified into a subgroup of glycoside hydrolase (GH) family 1 with β-Klotho, LPH (lactase-phlorizin hydrolase), KLrP, and CBG (cytosolic β-glucosidase) [23]. A couple of crystal structures of mammalian β-glycosidases of the GH family 1 have been solved [26]. Both the structures of KLrP and CBG consist of a (β/α) 8 TIM barrel of β-strands, in which the acid/base catalyst is located on strand 4 and nucleophilic glutamate is on strand 7. The cavity of the active center was occupied with lipids or steroids. Based on structural studies of GH family 1, it is possible that α-Klotho also recognizes some small molecules, such as sugars and lipids [26].

KLrP (klotho-related protein) is the closest homologue of α-Klotho and β-Klotho proteins (fig. 1). The crystal of KLrP was formed with a glucose-ceramide, which was docked at its active center. Moreover, two lipids were also contained in the crystal. The mutated KLrP that the acid/base glutamate was replaced to glutamine with was crystallized with glucose. Glucose was bound at the active center in the crystal as well [26]. Any crystal structures of α-Klotho or β-Klotho have not yet been solved. It is expected that determination of the structure will provide information about how α-Klotho specifically recognizes glucuronide and steroids.

Expression and Distribution

In the original report of the murine α-klotho gene, the expression pattern of the transcripts of the gene was analyzed via Northern blotting, RT-PCR, and in situ hybridization. The highest expression was detected in the distal convoluted tubule (DCT) of the kidney and in the choroid plexus of the brain [16].

Protein expression in mice was confirmed with anti-α-Klotho antibodies [19]. At least two types of α-Klotho protein were found, namely the membrane-bound type and the secreted one. The expression of membrane-bound α-Klotho is restricted in the DCT of the kidney, in the choroid plexus of the brain where mineral ions are actively transported and, of particular interest, in the parathyroid glands where PTH is synthesized to be secreted (fig. 2). Recently, α-Klotho expression was confirmed in syncytiotrophoblasts in normal term placenta, where nutrients are exchanged between mother and fetus and where progesterone is secreted [27]. Therefore, it is noticeable that α-Klotho is expressed in tissues where mineral ions are actively regulated and where hormones are secreted. On the other hand, secreted α-Klotho was

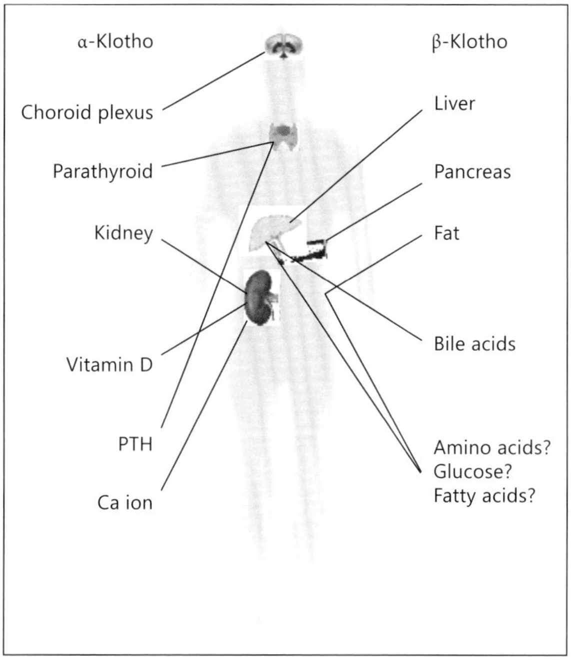

Fig. 2. A Schematic diagram summarizing expressions and functions of α-klotho and β-klotho.

detected not only in plasma, but also in cerebrospinal fluid (CSF) and urine [19].

We investigated the subcellular distribution of the α-Klotho protein using extracted choroid plexus in mice. In ex vivo studies, almost all α-Klotho protein was localized in intracellular compartments, but the amount of α-Klotho on the tissue surface was slight. Low extracellular calcium concentrations (~0.8 mM) induced the secretion of α-Klotho from the choroid plexus, the parathyroid glands, and the kidney [28, 29]. Secretion was reconstructed in α-Klotho expressing cultured HeLa cells [28]. With these observations, it is thought that secreted α-Klotho is a cleaved product of the membrane bound one induced by some extracellular stimuli.

Cleavage and Secretion

Based on the knowledge that α-Klotho is secreted with some extracellular stimuli, it has been hypothesized that secreted α-Klotho works as a hormone [30]. At this time, proposed stimuli which induce α-Klotho secretion are decreases in extracellular Ca concentrations or addition of insulin [4]. In ex vivo experiments, incubation or infusion with a buffer containing low Ca concentrations induced α-Klotho secretion from the kidney, choroid plexus, and parathyroid gland. Of note, the calcium-sensing receptor senses extracellular high calcium concentrations (~1.3 mM); meanwhile, any sensing machineries for monitoring extracellular low Ca concentrations (~0.8 mM) should exist but have not been identified [28]. On the other hand, in cultured cells, insulin induced α-Klotho secretion, which is caused by activation of a disintegrin and metalloproteases (ADAMs) 10/17 and β-amyloid precursor protein cleaving enzyme (BACE) 1 proteases (known as α-secretases and β-secretase, respectively) [31]. However, due to a lack of in vivo evidence, it has not been clarified as to which tissues are responsible for the insulin signal to release α-Klotho.

As mentioned above, α-Klotho secretion is reconstructed in cultured cells. Cells expressing the exogenous membrane-bound α-Klotho produced secreted α-Klotho accompanied with intracellular calcium influx within a couple of minutes. Secretion was not observed when the calcium influx was inhibited via chelating [28]. Low concentrations (2 mM) of a chelator in a cell-cultured medium, which does not affect protease activity, inhibited secretion. Therefore, it is not plausible that any proteases process membrane bound α-Klotho at the cell surface. Instead, it is preferable that vesicles containing previously cleaved α-Klotho are secreted by extracellular stimuli. Since any stimuli which induce Ca influx can induce α-Klotho secretion in cultured cell based assays, the physiological roles of α-Klotho secretion should be discussed with a data set of in vivo experiments [4].

As an Enzyme

Since α-klotho shares homology with β-glucosidase and lactase-phlorizin hydrolase, we assayed enzymatic activity and specificity of recombinant α-Klotho toward monosaccharide [32]. Purified α-Klotho specifically hydrolyzed glucuronic acid, but activity was quite low compared to that of natural β-glucuronidase. Low activity was explained by the fact that amino acid residues which are indispensable for the enzymatic reaction are not conserved in α-klotho. Low, but specific β-glucuronidase activity was inhibited in the presence of some glucuronized steroids. In addition, β-glucuronidase activity was highest in acidic conditions, but was lost at a neutral pH. Therefore, at that time, it was suggested that some glucuronized steroids or small molecules may be the substrates of α-Klotho in an acidic condition [32].

Interacting Partners

Several proteins to which α-Klotho interacts have been reported. In this section, we will introduce these studies and review them, with a focus on molecular aspects, e.g. (1) calculated binding constants are compared to physiological concentrations, (2) locations where molecular interactions occur are consistent to their tissue- and subcellular-distributions, and (3) conclusions led from only the results of in vitro or cultured cell based assays are adequate. The proposed manners of action can be categorized by the functions of α-Klotho: (a) as a component, α-Klotho directly or indirectly associates with partner proteins, or (b) as an enzyme, α-Klotho hydrolyzes a carbohydrate moiety of glycosylated proteins. The locations where these interactions occur can be divided into three spaces: (1) in intracellular vesicles, (2) on the cell surface, or (3) in body fluids such as blood or urine.

FGF23/FGFR1

Urakawa et al. [33] reported that α-Klotho directly binds to both FGF23 and FGFR1. Binding has been confirmed via in vitro assays, in cultured cell-based assays [34–36]. Later, these interactions were confirmed via in vivo experiments using mice [37]. It is thought that binding occurs at the basolateral membrane (blood side) surface of DCT cells of the kidney [38]. Comparing results that plasma concentrations of FGF23 are estimated below 10 pM [39], the binding constant of α-Klotho and FGF23 is still higher (3–5 nM) [Maeda et al., unpubl. data]. Therefore, there would be some other mechanisms for accumulating circulating FGF23 into the kidney. The physiological effects of the FGF23 signal (see below) are observed within the proximal convoluted tubule (PCT) cells, meanwhile expression of the α-klotho transcript is mostly observed in DCT cells, but scarcely in PCT cells [16].

As one of the physiological roles of the FGF23 signal for parathyroid glands (PT), it has been proposed that the signal suppresses synthesis of PTH [40]. However, there has been contradicting evidence that patients with chronic kidney disease suffer from hyperparathyroidism regardless of their excessive circulating concentrations of FGF23 [29]. Furthermore, only an increase in calcium ion concentration, not administration of vitamin D, PTH, Pi, or FGF23, suppressed α-klotho mRNA transcription in cultured parathyroid cells [29]. Indeed, in a wide human study, only serum calcium concentrations, not phosphate ions or PTH levels, were significantly correlated with α-klotho mRNA levels of parathyroid glands [41]. Therefore, these data do not support the idea that FGF23 directly influences parathyroid gland cells to suppress PTH synthesis.

α-Klotho is also expressed in the choroid plexus of the brain. It should be of note that, in the choroid plexus of adult mice, only FGFR2 is expressed among

the α-Klotho receptor has not been identified, the affinity of the hetero complex could not be argued. The administrated concentration (400 pM) where inhibitory effects of α-Klotho were observed was still higher than measured concentrations of α-Klotho in human plasma (~10 pM) [27, 39, 56]. The biological activity of the recombinant α-Klotho protein prepared from insect cells was not validated in this study. It has not been indicated which organs are targets of the secreted α-Klotho protein. Possible effects of α-Klotho on suppression of WNT and insulin signaling are elaborated below.

VEGF Receptor 2 and TRPC1

It has been reported that α-Klotho affects vascular calcium permeability via binding to vascular endothelial growth factor (VEGF) receptor 2 and transient-receptor potential canonical calcium channel 1 (TRPC1) on vascular endothelium [57]. The interaction was detected via immune-affinity assays and FRET microscopy analyses in primary cultured human umbilical vein endothelial cells and COS cells. In this report, only the C-terminal KL2 domain of α-Klotho was responsible for the interaction. Since mRNA levels of α-klotho in the vein were not detectable, secreted α-Klotho was supposed to affect vascular calcium permeability.

Although the binding assay was performed with conditioned medium containing secreted α-Klotho, the concentration of α-Klotho in the medium (~250 ng/ml) was still 250 times higher than that in human plasma (~1 ng/ml). There has been no in vivo evidence that α-Klotho interacts with VEGF receptor 2 or TRPC1. Additionally, it has not been shown if secreted α-Klotho accumulates on vascular endothelia.

Functions of α-Klotho

In this section, we introduce the recently proposed functions of α-Klotho according to its binding partners. These proposed functions will be reviewed mainly with data sets of the phenotypes of genetically engineered mice, results of in vivo experiments, and significances from wide human studies.

A Coreceptor for the FGF23 Signal

One of the most widely accepted functions of α-Klotho is as a coreceptor for the FGF23 signal in the kidney [33]. The overall phenotypes of both α-klotho knockout mice and fgf23 knockout mice are similar, especially in excessive vitamin D synthesis [58–61]. Vitamin D is activated by 25-hydroxyvitamin D_3-1α-hydroxylase (1α(OH)ase) in PCT cells of the kidney [62, 63]. Expression levels of 1α(OH)ase are abnormally high in both mice [17]. When both knockout mice were fed a vitamin D-deficient diet, overall observed disorders were ameliorated. Moreover, when both mice were crossed with 1α(OH)ase knockout mice, excessive vitamin D synthesis was rescued [17]. Therefore, it has been established that α-Klotho functions as a coreceptor for the FGF23 signal to suppress 1α(OH)ase expression in PCT cells.

A couple of questions remain. As mentioned above, whereas the effects of the FGF23 signal are observed in PCT cells, expression of α-Klotho is limited in DCT cells. A putative signal mediator from DCT cells to PCT cells has been proposed, but has not yet been identified [64, 65]. In the original paper of the α-klotho gene, phenotypes of α-klotho hypomorph mice were rescued by crossing with α-klotho transgenic mice, in which the exogenously introduced α-klotho gene was not expressed within the whole kidney [16]. Therefore, it may be hypothesized that the secreted α-Klotho protein binds to the PCT cell surface to input the FGF23 signal directly into PCT cells. One study has been reported where phosphorylation of ERK, which is caused by the FGF23 signal, was seen in a region of the kidney where α-Klotho protein is stained with an anti-α-Klotho antibody. The region was estimated as DCT cells, but was not fully demonstrated since any marker proteins were not investigated concomitantly [38, 66]. On the other hand, it is possible that, under some circumstances, the expression of α-Klotho is induced in PCT cells to input the FGF23 signal directly into cells.

Beyond DCT cells of the kidney, α-Klotho is expressed in the choroid plexus of the brain, PT, and the placenta. As mentioned above, any effects of the FGF23 signal for PT have been proposed, but have not yet been confirmed [41]. One of the proposed effects of the FGF23 signal in the PT is to suppress PT generation; however, the PT of FGF23 knockout mice are still small. There is evidence to suggest that patients of chronic kidney disease often generate hyperparathyroidism despite high FGF23 circulating levels [41]. Moreover, FGF23-dependent functions of α-Klotho in the choroid plexus of the brain or in the placenta have never been proposed.

Several slight, but significant different phenotypes between α-Klotho and FGF23 mice have been reported [60]. One of those is osteogenesis [67]. Bone formation deficiency is more prevalent in α-klotho knockout mice than in FGF23 mice. These data indicate that there are some FGF23-independent α-Klotho functions on PTH secretion and on regulating calcium and phosphate ion transport.

A Recruitor of the Na^+, K^+-ATPase Complex

Beyond the suppression of vitamin D synthesis, α-Klotho directly regulates calcium concentrations in some body fluids. It has been proposed that α-Klotho participates in calcium ion transport in the choroid plexus of the brain and in

DCT of the kidney. Furthermore, PTH secretion from the PT is also dependent on α-Klotho [28]. When calcium ion concentrations decrease in the circulation or in the CSF, the Na^+,K^+-ATPase complex comes up to the cell surface in an α-Klotho-dependent manner. Consequently, the surface amount of the Na^+,K^+-ATPase complex will be increased to make a steeper sodium gradient. This additively formed sodium gradient will be used for calcium ion transport via the NCX (sodium and calcium exchanger) in the choroid plexus and DCT cells. Moreover, at the PT, the sodium gradient is necessary for PTH secretion [28, 68–70].

While overall phenotypes of α-klotho knockout mice and fgf23 knockout mice are quite similar due to excessive vitamin D synthesis, phenotypes related to Na^+,K^+-ATPase recruitment are relatively slight and limited. Phenotypes observed in only α-klotho knockout mice, but not fgf23 knockout mice, should not be FGF23-dependent but Na^+,K^+-ATPase complex-dependent.

Comparing the phenotypes of α-klotho and fgf23 knockout mice, differences are observed in a PTH-dependent phenomenon. It is of note that abnormal bone formation, which is independent of the FGF23 signal, is still observed in α-klotho knockout mice [59, 67, 71–73]. Moreover, impaired PTH secretion and reduced calcium ion concentrations in the CSF are FGF23-independent phenotypes of α-klotho knockout mice [28].

It is of interest that the expression levels of α-klotho are negatively correlated with serum calcium concentrations, but not with serum phosphate ions, PTH, vitamin D, or FGF23 levels [41]. Moreover, the expression of α-klotho is decreased only with administration of calcium ions on cultured PT cells [29]. These observations also support the idea that α-Klotho is a component of sensing and regulating machinery of extracellular calcium ion concentrations.

Accumulating TRPV5 on the Cell Surface

The TRPV5 channel is necessary for calcium ion reabsorption in the kidney [74–76]. Whereas the sodium gradient made by the Na^+,K^+-ATPase complex is used as the driving force for drawing calcium ions from DCT cells into the blood [11], it has been proposed that α-Klotho activates the TRPV5 channel to uptake calcium ions from urine to DCT cells [50]. It was thought that the amount of TRPV5 channels at the apical membrane is under the control of α-Klotho, of which it is secreted in the urine, via hydrolyzing a glucuronyl moiety on the TRPV5 channel [50].

The working model that α-Klotho regulates TRPV5-dependent calcium ion reabsorption is of interest, but several questions remain to be addressed. There has been no evidence about the regulated secretion of α-Klotho itself from DCT cells into urine. Indeed, α-Klotho and the TRPV5 channel are both expressed in

DCT cells; however, secretion of α-Klotho is only observed at the basal membrane (blood side) [28]. It is possible that secreted α-Klotho in urine is just the consequence of passive excretion from plasma [77].

Furthermore, the effect of α-Klotho-dependent TRPV5 channel activation was observed in several hours (maximum at 16 h) after the addition of α-Klotho to cultured cells. It is a rather long period for regulating calcium ion reabsorption in DCT cells of the kidney.

Besides recent work showing that any glucuronyl moieties were not observed on the TRPV5 channel, other work proposed that α-Klotho hydrolyzes a sialic acid moiety instead of a glucuronyl moiety on the TRPV5 channel [51, 52]. When α-Klotho hydrolyzed a terminal sialic acid on the TRPV5 channel, an unmasked lactose carbohydrate would be captured by Galectin-1 and, subsequently, the channel would be retained on the cell surface [52]. However, any sialidase activities of α-Klotho have not been demonstrated and, of note, galectin-1 is expressed at the medullar interstitium (blood side of the DCT), but not at the apical membrane (urine side) of DCT cells [78]. Moreover, it has been reported that galectin-1 knockout mice do not express any phenotypes related to calcium homeostatic disorders [79–81].

TRPV5 and TRPV6 channels are expressed anywhere α-Klotho is expressed. It is of particular interest that the expression of TRPV5/6 channels and α-Klotho are both observed even in the trophoblast layer of the placenta which actively transports calcium ions between mother and fetus [27, 82]. These expression patterns suggest that both TRPV5/6 channels and α-Klotho cooperatively transport calcium ions. Therefore, it may be plausible that TRPV5/6 channels function for the intake of calcium ions from extracellular fluids and then calcium ions are excluded from cells using the sodium ion gradient made by the α-Klotho and Na^+,K^+-ATPase complex [83, 84].

Suppressor for the WNT Signal

Beyond the regulation of mineral homeostasis, it has been proposed that α-Klotho functions as a suppressor of the Wnt signal of stem cells [30]. When α-Klotho, of which the secreted type in blood is supposed, binds to Wnt, the Wnt signal that leads stem cell apoptosis would be suppressed. Therefore, the authors concluded that, in α-Klotho deficient mice, the over activated Wnt signal consequently reduces stem cell numbers [30].

The proposed working hypothesis should be reconsidered since it cannot explain the facts that a reduced vitamin D diet rescued the overall phenotypes of α-klotho knockout mice. Moreover, it should be explained why and how α-klotho knockout mice with reduced stem cell generation expressed such mineral homeostatic disorders.

A Suppressor for the Insulin/IGF Signal

It has been widely believed that suppression of insulin and the IGF signal would give mice a longer life [85–96]. In order to explain why the α-klotho gene is responsible for 'longevity', a hypothesis was proposed that secreted α-Klotho suppresses insulin and the IGF signal [97]. In this report, the authors concluded that a signal mediated by a putative α-Klotho receptor suppresses the insulin/IGF signal.

However, any target organs of secreted α-Klotho have not been indicated. Although the suppressing effects were observed in cultured cells, any receptors for α-Klotho have never been proposed. Actually, the α-Klotho-mediated suppression of the IGF signal has not been reproduced in any cultured cell based assays [98]. Therefore, the proposed hypothesis that α-Klotho is a suppressor for insulin and the IGF signal should be reconsidered.

Since reducing oxidative stresses are widely acknowledged ways to extend life span [99], it has been hypothesized that secreted α-Klotho also scavenges reactive oxygen species (ROS) [100]. Meanwhile, it has also been reported that secreted α-Klotho suppresses the TGF-β_1 signal [101]. However, in α-klotho knockout mice, an increase in any ROS or any overactivated TGF-β_1 signals has never been observed. These hypotheses cannot explain why α-klotho knockout mice could be rescued by a vitamin D-deficient diet [17]. Wide genetic studies did not identify any relationships between α-klotho gene polymorphisms and scavenging ability for ROS [102]. Finally, researchers have attempted to connect aging and phosphate regulation [103–107]; however, any relationships have never been demonstrated so far.

Accumulating TRPC1 on the Cell Surface

To explain the phenotype of arteriosclerosis seen in α-klotho knockout mice, it was proposed that secreted α-Klotho maintains vascular permeability via activating the TRPC1 channel on the vascular epidermis [57]. The working hypothesis is that secreted α-Klotho interacts with the VEGF receptor to increase the surface amount of TRPC1 channels via reducing endocytosis. The working model is similar to that of the TRPV5 channel in DCT cells of the kidney.

Unfortunately, the working hypothesis lacks consideration of the effects of vitamin D. It should be of note that α-klotho-deficient mice synthesize excessive vitamin D. Since the reduced vitamin D diet ameliorated arteriosclerosis in α-klotho knockout mice, the surface amount of TRPC1 channels may be under the control of vitamin D. Actually, it has been well known that expression of the TRPV6 channel is directly regulated to vitamin D [108].

Clinical Importance

Based on the acquired knowledge regarding the functions of α-Klotho, there have been attempts to diagnose and cure mineral homeostatic disorders related to α-Klotho functions. In the last section, we summarize proceeding clinical trials and proposed clinical usage of which α-Klotho is targeted.

Rickets and Kidney Failure

Since FGF23 suppresses vitamin D in an α-Klotho-dependent manner, inhibition of the interaction of FGF23 to α-Klotho has been tried to reduce hypervitamin D synthesis. Due to excessive vitamin D levels, patients with overproduction of FGF23 suffer from bone loss. Neutralizing antibodies of FGF23 have been generated to cure patients suffering from rickets and osteoporosis [109–111].

As it is now, concentrations of α-Klotho and FGF23 can be measured via ELISA systems [56, 112–114]. With wide clinical studies using ELISA systems, it has been proposed that patients out of the range of standard concentrations of α-Klotho or FGF23 are suspected of suffering from kidney failure, tumor formation, or a reduced number of osteocytes [39, 115]. Indeed, several patients suffer from rickets caused by a genetic α-klotho mutation [35, 116].

Hyper Parathyroidism

A large number of patients with kidney failure often suffer from hyperparathyroid glands, which have to be surgically removed. In these patients, overproduction of PTH accelerates bone loss. One agonist of the calcium-sensing receptor has been used for symptomatic therapy by reducing PTH secretion. As α-Klotho accelerates PTH secretion according to Na^+,K^+-ATPase activation in parathyroid glands, it is thought that suppression of α-Klotho-dependent Na^+,K^+-ATPase activation would reduce excessive secretion of PTH. Originally, Brown et al. [117] demonstrated that digitalis, ouabain, which inhibits Na^+,K^+-ATPase activity, suppresses PTH secretion. However, due to its strong side effects, any clinical applications have not been tried [117, 118]. Therefore, it is expected that binding of α-Klotho to the Na^+,K^+-ATPase complex could be one of the targets for curing patients of hyperparathyroidism.

Others

Enzymatic inhibitors of α-Klotho are potential suppressors for the TRPV5 channel [50]. However, any clinical attempts using these compounds have not been proposed. Meanwhile we have also argued that α-Klotho maintains mineral homeostasis of the CSF, which is quite important for normal neural activities [28]. Some mineral disorders of the CSF cause neuronal dysfunctions, such as epilepsy or paralysis. Indeed, we observed that α-klotho knockout mice were susceptible to epilepsy. Some neural dysfunctions would be cured by controlling α-Klotho activities to maintain mineral homeostasis in the CSF.

If secreted, α-Klotho functions as a hormone to suppress the Wnt signal, insulin/IGF signal, TGF-β_1 signal, ROS synthesis, or TRPC1 channel activation, though more studies are needed to confirm these proposed hypotheses, administration of a recombinant secreted α-Klotho itself may be useful as a drug. In these cases, preparation of the recombinant α-Klotho protein should be validated since, in our experiments, any recombinant α-Klotho protein produced in insect cells was inactive. Actually, any attempts for clinical use of secreted α-Klotho protein have not been tried so far.

Conclusion

Although it is well known that α-Klotho regulates mineral homeostasis of phosphate and calcium ions, the precise molecular mechanisms remain to be revealed. One of the most important studies is to determine the structure of α-Klotho, which would explain why α-Klotho interacts with such a vast repertoire of proteins. Studies about β-klotho would also help us to understand the physiological roles of Klotho family proteins.

Besides vitamin D and PTH, which regulate mineral homeostasis as hormones, α-Klotho is a new member for mineral regulating factors [119, 120]. It is obvious that all animals, including humans, maintain every kind of homeostasis, including glucose, lipids, and amino acids. Klotho family proteins probably regulate these metabolites beyond hormones (fig. 2).

References

1 Weindruch R, Walford RL, Fligiel S, Guthrie D: The retardation of aging in mice by dietary restriction: longevity, cancer, immunity and lifetime energy intake. J Nutr 1986;116:641–654.

2 Finkel T, Holbrook NJ: Oxidants, oxidative stress and the biology of ageing. Nature 2000; 408:239–247.

3 Finch CE, Ruvkun G: The genetics of aging. Annu Rev Genomics Hum Genet 2001;2: 435–462.

4 Chen CD, Podvin S, Gillespie E, Leeman SE, Abraham CR: Insulin stimulates the cleavage and release of the extracellular domain of klotho by adam10 and adam17. Proc Natl Acad Sci USA 2007;104:19796–19801.

5 Sinclair DA, Oberdoerffer P: The ageing epigenome: damaged beyond repair? Ageing Res Rev 2009;8:189–198.

6 Haigis MC, Sinclair DA: Mammalian sirtuins: biological insights and disease relevance. Annu Rev Pathol;5:253–295.

7 Baur JA, Chen D, Chini EN, Chua K, Cohen HY, de Cabo R, Deng C, Dimmeler S, Gius D, Guarente LP, Helfand SL, Imai S, Itoh H, Kadowaki T, Koya D, Leeuwenburgh C, McBurney M, Nabeshima Y, Neri C, Oberdoerffer P, Pestell RG, Rogina B, Sadoshima J, Sartorelli V, Serrano M, Sinclair DA, Steegborn C, Tatar M, Tissenbaum HA, Tong Q, Tsubota K, Vaquero A, Verdin E: Dietary restriction: standing up for sirtuins. Science 2010;329:1012–1013;author reply 1013–1014.
8 Boskey AL, Coleman R: Aging and bone. J Dent Res 2010;89:1333–1348.
9 Esposito C, Dal Canton A: Functional changes in the aging kidney. J Nephrol 2010; 23(suppl 15):S41–S45.
10 Nabeshima Y, Imura H: Alpha-klotho: a regulator that integrates calcium homeostasis. Am J Nephrol 2008;28:455–464.
11 Nabeshima Y: Discovery of alpha-klotho unveiled new insights into calcium and phosphate homeostasis. Proc Jpn Acad Ser B Phys Biol Sci 2009;85:125–141.
12 Sugano Y, Lardelli M: Identification and expression analysis of the zebrafish orthologue of klotho. Dev Genes Evol 2011;221:179–186.
13 Okabe M, Graham A: The origin of the parathyroid gland. Proc Natl Acad Sci USA 2004; 101:17716–17719.
14 Polanska UM, Edwards E, Fernig DG, Kinnunen TK: The cooperation of FGF receptor and klotho is involved in excretory canal development and regulation of metabolic homeostasis in *Caenorhabditis elegans*. J Biol Chem 2011;286:5657–5666.
15 Chateau MT, Araiz C, Descamps S, Galas S: Klotho interferes with a novel FGF-signalling pathway and insulin/IgF-like signalling to improve longevity and stress resistance in *Caenorhabditis elegans*. Aging (Albany NY) 2010;2:567–581.
16 Kuro-o M, Matsumura Y, Aizawa H, Kawaguchi H, Suga T, Utsugi T, Ohyama Y, Kurabayashi M, Kaname T, Kume E, Iwasaki H, Iida A, Shiraki-Iida T, Nishikawa S, Nagai R, Nabeshima YI: Mutation of the mouse klotho gene leads to a syndrome resembling ageing. Nature 1997;390:45–51.
17 Tsujikawa H, Kurotaki Y, Fujimori T, Fukuda K, Nabeshima Y: Klotho, a gene related to a syndrome resembling human premature aging, functions in a negative regulatory circuit of vitamin D endocrine system. Mol Endocrinol 2003;17:2393–2403.
18 Morishita K, Shirai A, Kubota M, Katakura Y, Nabeshima Y, Takeshige K, Kamiya T: The progression of aging in klotho mutant mice can be modified by dietary phosphorus and zinc. J Nutr 2001;131:3182–3188.
19 Imura A, Iwano A, Tohyama O, Tsuji Y, Nozaki K, Hashimoto N, Fujimori T, Nabeshima Y: Secreted klotho protein in sera and CSF: Implication for post-translational cleavage in release of klotho protein from cell membrane. FEBS Lett 2004;565:143–147.
20 Matsumura Y, Aizawa H, Shiraki-Iida T, Nagai R, Kuro-o M, Nabeshima Y: Identification of the human klotho gene and its two transcripts encoding membrane and secreted klotho protein. Biochem Biophys Res Commun 1998;242:626–630.
21 Shiraki-Iida T, Aizawa H, Matsumura Y, Sekine S, Iida A, Anazawa H, Nagai R, Kuro-o M, Nabeshima Y: Structure of the mouse klotho gene and its two transcripts encoding membrane and secreted protein. FEBS Lett 1998;424:6–10.
22 Wacker H, Keller P, Falchetto R, Legler G, Semenza G: Location of the two catalytic sites in intestinal lactase-phlorizin hydrolase. Comparison with sucrase-isomaltase and with other glycosidases, the membrane anchor of lactase-phlorizin hydrolase. J Biol Chem 1992;267:18744–18752.
23 Coutinho PM, Henrissat B: Life with no sugars? J Mol Microbiol Biotechnol 1999;1:307–308.
24 Ito S, Kinoshita S, Shiraishi N, Nakagawa S, Sekine S, Fujimori T, Nabeshima YI: Molecular cloning and expression analyses of mouse betaklotho, which encodes a novel klotho family protein. Mech Dev 2000;98:115–119.
25 Ito S, Fujimori T, Furuya A, Satoh J, Nabeshima Y: Impaired negative feedback suppression of bile acid synthesis in mice lacking betaklotho. J Clin Invest 2005;115:2202–2208.
26 Noguchi J, Hayashi Y, Baba Y, Okino N, Kimura M, Ito M, Kakuta Y: Crystal structure of the covalent intermediate of human cytosolic beta-glucosidase. Biochem Biophys Res Commun 2008;374:549–552.
27 Ohata Y, Arahori H, Namba N, Kitaoka T, Hirai H, Wada K, Nakayama M, Michigami T, Imura A, Nabeshima Y, Yamazaki Y, Ozono K: Circulating levels of soluble alpha-klotho are markedly elevated in human umbilical cord blood. J Clin Endocrinol Metab 2011;96:E943–E947.

28 Imura A, Tsuji Y, Murata M, Maeda R, Kubota K, Iwano A, Obuse C, Togashi K, Tominaga M, Kita N, Tomiyama K, Iijima J, Nabeshima Y, Fujioka M, Asato R, Tanaka S, Kojima K, Ito J, Nozaki K, Hashimoto N, Ito T, Nishio T, Uchiyama T, Fujimori T: Alpha-klotho as a regulator of calcium homeostasis. Science 2007;316:1615–1618.
29 Krajisnik T, Olauson H, Mirza MA, Hellman P, Akerstrom G, Westin G, Larsson TE, Bjorklund P: Parathyroid klotho and FGF-receptor 1 expression decline with renal function in hyperparathyroid patients with chronic kidney disease and kidney transplant recipients. Kidney Int 2010;78:1024–1032.
30 Liu H, Fergusson MM, Castilho RM, Liu J, Cao L, Chen J, Malide D, Rovira II, Schimel D, Kuo CJ, Gutkind JS, Hwang PM, Finkel T: Augmented Wnt signaling in a mammalian model of accelerated aging. Science 2007;317: 803–806.
31 Bloch L, Sineshchekova O, Reichenbach D, Reiss K, Saftig P, Kuro-o M, Kaether C: Klotho is a substrate for alpha-, beta- and gamma-secretase. FEBS Lett 2009;583:3221–3224.
32 Tohyama O, Imura A, Iwano A, Freund JN, Henrissat B, Fujimori T, Nabeshima Y: Klotho is a novel beta-glucuronidase capable of hydrolyzing steroid beta-glucuronides. J Biol Chem 2004;279:9777–9784.
33 Urakawa I, Yamazaki Y, Shimada T, Iijima K, Hasegawa H, Okawa K, Fujita T, Fukumoto S, Yamashita T: Klotho converts canonical FGF receptor into a specific receptor for FGF23. Nature 2006;444:770–774.
34 Kurosu H, Ogawa Y, Miyoshi M, Yamamoto M, Nandi A, Rosenblatt KP, Baum MG, Schiavi S, Hu MC, Moe OW, Kuro-o M: Regulation of fibroblast growth factor-23 signaling by klotho. J Biol Chem 2006;281:6120–6123.
35 Ichikawa S, Imel EA, Kreiter ML, Yu X, Mackenzie DS, Sorenson AH, Goetz R, Mohammadi M, White KE, Econs MJ: A homozygous missense mutation in human klotho causes severe tumoral calcinosis. J Musculoskelet Neuronal Interact 2007;7:318–319.
36 Goetz R, Nakada Y, Hu MC, Kurosu H, Wang L, Nakatani T, Shi M, Eliseenkova AV, Razzaque MS, Moe OW, Kuro-o M, Mohammadi M: Isolated c-terminal tail of FGF23 alleviates hypophosphatemia by inhibiting FGF23-FGFr-klotho complex formation. Proc Natl Acad Sci USA 2009;107:407–412.
37 Tomiyama KI, Maeda R, Urakawa I, Yamazaki Y, Tanaka T, Ito S, Nabeshima Y, Tomita T, Odori S, Hosoda K, Nakao K, Imura A, Nabeshima YI: Relevant use of klotho in FGF19 subfamily signaling system in vivo. Proc Natl Acad Sci USA 2010;107:1666–1671.
38 Farrow EG, Davis SI, Summers LJ, White KE: Initial FGF23-mediated signaling occurs in the distal convoluted tubule. J Am Soc Nephrol 2009;20:955–960.
39 Sugiura H, Tsuchiya K, Nitta K: Circulating levels of soluble alpha-klotho in patients with chronic kidney disease. Clin Exp Nephrol 2011;15:795–796.
40 Ben-Dov IZ, Galitzer H, Lavi-Moshayoff V, Goetz R, Kuro-o M, Mohammadi M, Sirkis R, Naveh-Many T, Silver J: The parathyroid is a target organ for FGF23 in rats. J Clin Invest 2007;117:4003–4008.
41 Bjorklund P, Krajisnik T, Akerstrom G, Westin G, Larsson TE: Type I membrane klotho expression is decreased and inversely correlated to serum calcium in primary hyperparathyroidism. J Clin Endocrinol Metab 2008;93: 4152–4157.
42 Yazaki N, Hosoi Y, Kawabata K, Miyake A, Minami M, Satoh M, Ohta M, Kawasaki T, Itoh N: Differential expression patterns of mRNAs for members of the fibroblast growth factor receptor family, FGFr-1-FGFr-4, in rat brain. J Neurosci Res 1994;37:445–452.
43 Reid S, Ferretti P: Differential expression of fibroblast growth factor receptors in the developing murine choroid plexus. Brain Res Dev Brain Res 2003;141:15–24.
44 King GD, Rosene DL, Abraham CR: Promoter methylation and age-related downregulation of klotho in rhesus monkey. Age (Dordr) 2012;34:1405-1419.
45 Sopjani M, Alesutan I, Dermaku-Sopjani M, Gu S, Zelenak C, Munoz C, Velic A, Foller M, Rosenblatt KP, Kuro-o M, Lang F: Regulation of the Na^{+}/K^{+} ATPase by klotho. FEBS Lett 2011;585:1759–1764.
46 Marsell R, Krajisnik T, Goransson H, Ohlsson C, Ljunggren O, Larsson TE, Jonsson KB: Gene expression analysis of kidneys from transgenic mice expressing fibroblast growth factor-23. Nephrol Dial Transplant 2008;23: 827–833.
47 Chow DC, Forte JG: Functional significance of the beta-subunit for heterodimeric p-type atpases. J Exp Biol 1995;198:1–17.

48 Geering K: The functional role of beta subunits in oligomeric p-type atpases. J Bioenerg Biomembr 2001;33:425–438.
49 Kaplan JH: Biochemistry of Na,K-ATPase. Annu Rev Biochem 2002;71:511–535.
50 Chang Q, Hoefs S, van der Kemp AW, Topala CN, Bindels RJ, Hoenderop JG: The beta-glucuronidase klotho hydrolyzes and activates the TRPV5 channel. Science 2005;310:490–493.
51 Morita I, Kizuka Y, Kakuda S, Oka S: Expression and function of the HNK-1 carbohydrate. J Biochem 2008;143:719–724.
52 Cha SK, Ortega B, Kurosu H, Rosenblatt KP, Kuro OM, Huang CL: Removal of sialic acid involving klotho causes cell-surface retention of TRPV5 channel via binding to galectin-1. Proc Natl Acad Sci USA 2008;105:9805–9810.
53 Medici D, Razzaque MS, Deluca S, Rector TL, Hou B, Kang K, Goetz R, Mohammadi M, Kuro OM, Olsen BR, Lanske B: FGF-23-klotho signaling stimulates proliferation and prevents vitamin D-induced apoptosis. J Cell Biol 2008;182:459–465.
54 Rando TA: Stem cells, ageing and the quest for immortality. Nature 2006;441:1080–1086.
55 Jones DL, Rando TA: Emerging models and paradigms for stem cell ageing. Nat Cell Biol 2011;13:506–512.
56 Yamazaki Y, Imura A, Urakawa I, Shimada T, Murakami J, Aono Y, Hasegawa H, Yamashita T, Nakatani K, Saito Y, Okamoto N, Kurumatani N, Namba N, Kitaoka T, Ozono K, Sakai T, Hataya H, Ichikawa S, Imel EA, Econs MJ, Nabeshima Y: Establishment of sandwich ELISA for soluble alpha-klotho measurement: age-dependent change of soluble alpha-klotho levels in healthy subjects. Biochem Biophys Res Commun 2010;398:513–518.
57 Kusaba T, Okigaki M, Matui A, Murakami M, Ishikawa K, Kimura T, Sonomura K, Adachi Y, Shibuya M, Shirayama T, Tanda S, Hatta T, Sasaki S, Mori Y, Matsubara H: Klotho is associated with VEGF receptor-2 and the transient receptor potential canonical-1 Ca^{2+} channel to maintain endothelial integrity. Proc Natl Acad Sci USA 2010;107:19308–19313.
58 Shimada T, Kakitani M, Yamazaki Y, Hasegawa H, Takeuchi Y, Fujita T, Fukumoto S, Tomizuka K, Yamashita T: Targeted ablation of FGF23 demonstrates an essential physiological role of FGF23 in phosphate and vitamin D metabolism. J Clin Invest 2004;113:561–568.
59 Sitara D, Razzaque MS, St-Arnaud R, Huang W, Taguchi T, Erben RG, Lanske B: Genetic ablation of vitamin D activation pathway reverses biochemical and skeletal anomalies in FGF-23-null animals. Am J Pathol 2006;169:2161–2170.
60 Razzaque MS, Lanske B: Hypervitaminosis D and premature aging: lessons learned from FGF23 and klotho mutant mice. Trends Mol Med 2006;12:298–305.
61 Razzaque MS, Sitara D, Taguchi T, St-Arnaud R, Lanske B: Premature aging-like phenotype in fibroblast growth factor 23 null mice is a vitamin D-mediated process. FASEB J 2006;20:720–722.
62 Brown AJ, Ritter C, Slatopolsky E, Muralidharan KR, Okamura WH, Reddy GS: 1alpha,25-dihydroxy-3-epi-vitamin D_3, a natural metabolite of 1alpha,25-dihydroxyvitamin D_3, is a potent suppressor of parathyroid hormone secretion. J Cell Biochem 1999;73:106–113.
63 Hewison M, Zehnder D, Bland R, Stewart PM: 1alpha-Hydroxylase and the action of vitamin D. J Mol Endocrinol 2000;25:141–148.
64 Segawa H, Yamanaka S, Ohno Y, Onitsuka A, Shiozawa K, Aranami F, Furutani J, Tomoe Y, Ito M, Kuwahata M, Imura A, Nabeshima Y, Miyamoto K: Correlation between hyperphosphatemia and type II Na-Pi cotransporter activity in klotho mice. Am J Physiol Renal Physiol 2007;292:F769–F779.
65 Dermaku-Sopjani M, Sopjani M, Saxena A, Shojaiefard M, Bogatikov E, Alesutan I, Eichenmuller M, Lang F: Downregulation of NaPi-IIa and NaPi-IIb Na-coupled phosphate transporters by coexpression of klotho. Cell Physiol Biochem 2011;28:251–258.
66 Hu MC, Shi M, Zhang J, Pastor J, Nakatani T, Lanske B, Razzaque MS, Rosenblatt KP, Baum MG, Kuro-o M, Moe OW: Klotho: A novel phosphaturic substance acting as an autocrine enzyme in the renal proximal tubule. FASEB J 2010;24:3438–3450.
67 Wang Y, Sun Z: Current understanding of klotho. Ageing Res Rev 2009;8:43–51.
68 Strewler GJ: Untangling klotho's role in calcium homeostasis. Cell Metab 2007;6:93–95.
69 Razzaque MS, Lanske B: The emerging role of the fibroblast growth factor-23-klotho axis in renal regulation of phosphate homeostasis. J Endocrinol 2007;194:1–10.

70 Razzaque MS: Klotho and Na^+,K^+-ATPase activity: solving the calcium metabolism dilemma? Nephrol Dial Transplant 2008;23: 459–461.
71 Suzuki H, Amizuka N, Oda K, Noda M, Ohshima H, Maeda T: Histological and elemental analyses of impaired bone mineralization in klotho-deficient mice. J Anat 2008; 212:275–285.
72 Suzuki H, Amizuka N, Oda K, Noda M, Ohshima H, Maeda T: Involvement of the klotho protein in dentin formation and mineralization. Anat Rec (Hoboken) 2008;291: 183–190.
73 Yuan Q, Sato T, Densmore M, Saito H, Schuler C, Erben RG, Lanske B: FGF-23/ klotho signaling is not essential for the phosphaturic and anabolic functions of PTH. J Bone Miner Res 2011;26:2026–2035.
74 Hoenderop JG, van Leeuwen JP, van der Eerden BC, Kersten FF, van der Kemp AW, Merillat AM, Waarsing JH, Rossier BC, Vallon V, Hummler E, Bindels RJ: Renal Ca^{2+} wasting, hyperabsorption, and reduced bone thickness in mice lacking TRPV5. J Clin Invest 2003;112:1906–1914.
75 Clapham DE: TRP channels as cellular sensors. Nature 2003;426:517–524.
76 van Abel M, Hoenderop JG, Bindels RJ: The epithelial calcium channels TRPV5 and TRPV6: regulation and implications for disease. Naunyn Schmiedebergs Arch Pharmacol 2005;371:295–306.
77 Cha SK, Hu MC, Kurosu H, Kuro-o M, Moe O, Huang CL: Regulation of renal outer medullary potassium channel and renal K(+) excretion by klotho. Mol Pharmacol 2009;76: 38–46.
78 Vansthertem D, Cludts S, Nonclercq D, Gossiaux A, Saussez S, Legrand A, Gabius HJ, Toubeau G: Immunohistochemical localization of galectins-1 and -3 and monitoring of tissue galectin-binding sites during tubular regeneration after renal ischemia reperfusion in the rat. Histol Histopathol 2010;25:1417–1429.
79 Poirier F, Robertson EJ: Normal development of mice carrying a null mutation in the gene encoding the l14 S-type lectin. Development 1993;119:1229–1236.
80 McGraw J, Gaudet AD, Oschipok LW, Steeves JD, Poirier F, Tetzlaff W, Ramer MS: Altered primary afferent anatomy and reduced thermal sensitivity in mice lacking galectin-1. Pain 2005;114:7–18.
81 Case D, Irwin D, Ivester C, Harral J, Morris K, Imamura M, Roedersheimer M, Patterson A, Carr M, Hagen M, Saavedra M, Crossno J Jr, Young KA, Dempsey EC, Poirier F, West J, Majka S: Mice deficient in galectin-1 exhibit attenuated physiological responses to chronic hypoxia-induced pulmonary hypertension. Am J Physiol Lung Cell Mol Physiol 2007;292:L154–L164.
82 Stumpf T, Zhang Q, Hirnet D, Lewandrowski U, Sickmann A, Wissenbach U, Dorr J, Lohr C, Deitmer JW, Fecher-Trost C: The human TRPV6 channel protein is associated with cyclophilin B in human placenta. J Biol Chem 2008;283:18086–18098.
83 van der Eerden BC, Weissgerber P, Fratzl-Zelman N, Olausson J, Hoenderop JG, Schreuders-Koedam M, Eijken M, Roschger P, de Vries TJ, Chiba H, Klaushofer K, Flockerzi V, Bindels RJ, Freichel M, van Leeuwen JP: The transient receptor potential channel TRPV6 is dynamically expressed in bone cells but is not crucial for bone mineralization in mice. J Cell Physiol 2010;227:1951–1959.
84 Boros S, Bindels RJ, Hoenderop JG: Active Ca(2+) reabsorption in the connecting tubule. Pflügers Arch 2009;458:99–109.
85 Kenyon C, Chang J, Gensch E, Rudner A, Tabtiang R: A *C. elegans* mutant that lives twice as long as wild type. Nature 1993;366: 461–464.
86 Kloting N, Bluher M: Extended longevity and insulin signaling in adipose tissue. Exp Gerontol 2005;40:878–883.
87 Bluher M, Kahn BB, Kahn CR: Extended longevity in mice lacking the insulin receptor in adipose tissue. Science 2003;299:572–574.
88 Kenyon C: The first long-lived mutants: discovery of the insulin/IgF-1 pathway for ageing. Phil Trans R Soc Lond B Biol Sci 2011; 366:9–16.
89 Holzenberger M, Dupont J, Ducos B, Leneuve P, Geloen A, Even PC, Cervera P, Le Bouc Y: IgF-1 receptor regulates lifespan and resistance to oxidative stress in mice. Nature 2003;421:182–187.
90 Giannakou ME, Partridge L: Role of insulin-like signalling in *Drosophila* lifespan. Trends Biochem Sci 2007;32:180–188.
91 Hwangbo DS, Gershman B, Tu MP, Palmer M, Tatar M: *Drosophila* dFOXO controls lifespan and regulates insulin signalling in brain and fat body. Nature 2004;429:562–566.

92 Holzenberger M, Kappeler L, De Magalhaes Filho C: IgF-1 signaling and aging. Exp Gerontol 2004;39:1761–1764.
93 Clancy DJ, Gems D, Harshman LG, Oldham S, Stocker H, Hafen E, Leevers SJ, Partridge L: Extension of life-span by loss of CHICO, a *Drosophila* insulin receptor substrate protein. Science 2001;292:104–106.
94 Tatar M, Kopelman A, Epstein D, Tu MP, Yin CM, Garofalo RS: A mutant drosophila insulin receptor homolog that extends life-span and impairs neuroendocrine function. Science 2001;292:107–110.
95 Tatar M, Bartke A, Antebi A: The endocrine regulation of aging by insulin-like signals. Science 2003;299:1346–1351.
96 Barbieri M, Rizzo MR, Manzella D, Grella R, Ragno E, Carbonella M, Abbatecola AM, Paolisso G: Glucose regulation and oxidative stress in healthy centenarians. Exp Gerontol 2003;38:137–143.
97 Kurosu H, Yamamoto M, Clark JD, Pastor JV, Nandi A, Gurnani P, McGuinness OP, Chikuda H, Yamaguchi M, Kawaguchi H, Shimomura I, Takayama Y, Herz J, Kahn CR, Rosenblatt KP, Kuro-o M: Suppression of aging in mice by the hormone klotho. Science 2005;309:1829–1833.
98 Lorenzi O, Veyrat-Durebex C, Wollheim CB, Villemin P, Rohner-Jeanrenaud F, Zanchi A, Vischer UM: Evidence against a direct role of klotho in insulin resistance. Pflügers Arch 2010;459:465–473.
99 Schriner SE, Linford NJ, Martin GM, Treuting P, Ogburn CE, Emond M, Coskun PE, Ladiges W, Wolf N, Van Remmen H, Wallace DC, Rabinovitch PS: Extension of murine life span by overexpression of catalase targeted to mitochondria. Science 2005;308:1909–1911.
100 Yamamoto M, Clark JD, Pastor JV, Gurnani P, Nandi A, Kurosu H, Miyoshi M, Ogawa Y, Castrillon DH, Rosenblatt KP, Kuro-o M: Regulation of oxidative stress by the anti-aging hormone klotho. J Biol Chem 2005;280:38029–38034.
101 Doi S, Zou Y, Togao O, Pastor JV, John GB, Wang L, Shiizaki K, Gotschall R, Schiavi S, Yorioka N, Takahashi M, Boothman DA, Kuro-o M: Klotho inhibits transforming growth factor-beta1 (TGF-beta1) signaling and suppresses renal fibrosis and cancer metastasis in mice. J Biol Chem 2011;286:8655–8665.
102 Karsli Ceppioglu S, Yurdun T, Canbakan M: Assessment of matrix GLA protein, klotho gene polymorphisms, and oxidative stress in chronic kidney disease. Ren Fail 2011;33:866–874.
103 Kuro-o M: A potential link between phosphate and aging – lessons from klotho-deficient mice. Mech Ageing Dev 2010;131:270–275.
104 Kuro-o M: Klotho and aging. Biochim Biophys Acta 2009;1790:1049–1058.
105 Kuro-o M: Klotho in chronic kidney disease – what's new? Nephrol Dial Transplant 2009;24:1705–1708.
106 Ohnishi M, Razzaque MS: Dietary and genetic evidence for phosphate toxicity accelerating mammalian aging. FASEB J 2010;24:3562–3571.
107 John GB, Cheng CY, Kuro-o M: Role of klotho in aging, phosphate metabolism, and CKD. Am J Kidney Dis 2010;58:127–134.
108 Wood RJ, Tchack L, Taparia S: 1,25-Dihydroxyvitamin D_3 increases the expression of the CaT1 epithelial calcium channel in the Caco-2 human intestinal cell line. BMC Physiol 2001;1:11.
109 Shimada T, Mizutani S, Muto T, Yoneya T, Hino R, Takeda S, Takeuchi Y, Fujita T, Fukumoto S, Yamashita T: Cloning and characterization of FGF23 as a causative factor of tumor-induced osteomalacia. Proc Natl Acad Sci USA 2001;98:6500–6505.
110 Shimada T, Hasegawa H, Yamazaki Y, Muto T, Hino R, Takeuchi Y, Fujita T, Nakahara K, Fukumoto S, Yamashita T: FGF-23 is a potent regulator of vitamin d metabolism and phosphate homeostasis. J Bone Miner Res 2004;19:429–435.
111 Yamazaki Y, Tamada T, Kasai N, Urakawa I, Aono Y, Hasegawa H, Fujita T, Kuroki R, Yamashita T, Fukumoto S, Shimada T: Anti-FGF23 neutralizing antibodies show the physiological role and structural features of FGF23. J Bone Miner Res 2008;23:1509–1518.
112 Yamazaki Y, Okazaki R, Shibata M, Hasegawa Y, Satoh K, Tajima T, Takeuchi Y, Fujita T, Nakahara K, Yamashita T, Fukumoto S: Increased circulatory level of biologically active full-length FGF-23 in patients with hypophosphatemic rickets/osteomalacia. J Clin Endocrinol Metab 2002;87:4957–4960.

desperately needed for preserving kidney function and minimizing cardiovascular complications.

The *Klotho* gene was identified in 1997 when a locus was accidentally interrupted that resulted in a cluster of features resembling premature multiorgan failure [3, 4]. Klotho is expressed the highest in the kidney and, interestingly, the phenotype of the Klotho hypomorph or knock-out mice, have striking similarities with CKD patients such as ectopic soft tissue calcification, hyperphosphatemia, and high plasma FGF23 [5], suggesting that CKD might be a state of Klotho deficiency.

Klotho is a single-pass transmembrane protein and is expressed in multiple tissues but is particularly high in the kidney [4, 6]. In the mouse, rat, and human kidney, Klotho is strongly expressed in the distal convoluted tubules but also unequivocally found in the proximal convoluted tubule [7, 8] and perhaps in some inner medullary collecting cells although in lower levels compared to distal convoluted tubules [7, 9]. In addition to membrane Klotho, a secreted form of Klotho protein is generated from the *Klotho* gene through alternative splicing and directly released into extracellular compartment and present in multiple body fluids. A second form of soluble Klotho is ectodomain shedding of membrane Klotho from the cell surface under the effect of A Desintegrin And Metalloproteinase (ADAM)10 and ADAM17, which is stimulated by insulin and inhibited by metalloproteinase inhibitors [10]. Soluble Klotho protein is present in cerebrospinal fluid [11], blood [5, 11], and urine of mammals [5, 8].

Membrane-anchored and soluble Klotho proteins likely have distinct functions. Membrane Klotho forms a complex with fibroblast growth factor (FGF) receptors and functions as a coreceptor for FGF23, a bone-derived hormone that induces negative phosphate balance through promoting renal phosphate excretion and reducing serum levels of 1,25-dihydroxyvitamin D_3 (1,25-$(OH)_2VD_3$) [12]. Soluble Klotho functions as an endocrine factor, and plays important roles in a variety of processes including modulation of ion transport [8, 13], Wnt signal transduction [14], anti-renin-angiotensin system [15], antisenescence [16], and antioxidation [17]. In addition, Klotho suppresses 1α-hydroxylase in the kidney to regulate calcium metabolism [18] and participates in the regulation of PTH synthesis in the parathyroid gland by FGF23 [19].

Klotho Deficiency in Chronic Kidney Disease

When one compares the features of Klotho-deficient mice with manifestations of CKD, there are interesting similarities including hyperphosphatemia, high plasma FGF23, ectopic soft tissue calcification, and decreased Klotho in the

Table 1. Endocrine and renal Klotho deficiency in CKD in humans and in rodent models

Category of kidney disease	Cause or animal model	Renal Klotho protein		Renal Klotho mRNA		Blood Klotho assay	Urine Klotho assay	Human or rodent			Reference no.
		IB	IHC	NB	qPCR			M	R	H	
CKD	Npx+IRI	↓	↓	↓	↓	↓ by IP + IB	↓ by IB	√	√		5
CKD	5/6 Npx			↓					√		28
CKD	ICGN				↓						24
CKD	Npx in apo-E–/–	↓			↓	↓ by ELISA		√			24
Hypertension	SHR			↓					√		28
Hypertension	DOCA			↓					√		28
Hypertension	Ang II	↓	↓	↓					√		16
Hypertension	Ang II	↓		↓					√		67
Hypertension	SHR	↓	↓		↓				√		27
DM	OLETF			↓					√		28
DM	streptozotocin	↓						√			39
DM	*db/db*	↓	↓		↓			√			26
CKD	CGN	↓	↓	↓						√	2
CKD	DN	↓	↓	↓						√	21
Allograft	CGR	↓	↓	↓						√	21
CKD	N/A						↓ by IB			√	5
CKD	N/A					↑ by ELISA				√	35
Adults	community >65 years					↓ by ELISA with high death				√	33
Adults	community 24–102 years					↓ by ELISA with high CVD				√	34
Children/adults	normal					↓ by ELISA with age				√	32
Age/aged kidney	normal	↓						√			30

CGN = Chronic glomerulonephritis; CsA = cyclosporine A; CVD = cardiovascular disease; DM = Diabetes mellitus; DN = Diabetic nephropathy; DOCA-salt = deoxycorticosterone acetate and high salt intake (DOCA-Salt); FK = FK-506, or Tacrolimus; IB = Immunoblot; ICGN = ICR-derived spontaneous glomerulonephritis; IHC = immunohistochesimstry; IP = immunoprecipitation; NB = Northern blot; Npx = nephrectomy; OLETF = the Otsuka Long-Evans Tokushima fatty rat; RT = reverse transcription; SHR = spontaneous hypertension; qPCR = quantitative or real time PCR; CGR = chronic graft rejection.

blood and kidney, suggesting that CKD might be a state of Klotho deficiency. Experimental data and clinical findings have thus far supported this view (table 1) [17, 20–26].

Aging Kidney

Aging rodents (older than 2 years) have significantly higher serum creatinine (S_{Cr}) [27] and lower renal Klotho protein expression compared to young ones [28]. Notably, the decreased renal Klotho protein levels in aged rodents is associated with increased oxidative stress, proinflammatory cytokine production, and activation of endothelin signal transduction (table 1) [27].

There are numerous epidemiological studies showing association of *Klotho* gene or promoter polymorphisms with a variety of phenotypes in several ethic populations. A few are noteworthy. *Klotho* gene polymorphism F352V and/or C370S within exon 2 increases secreted levels and β-glucosidase activity of Klotho in vitro in cultured cells, suggesting that *Klotho* gene variants may influence the trafficking and catalytic activity of Klotho [29]. However, one epidemiological study showed that the heterozygous genotype (Klotho-*VS/WT*) but not the Klotho-*VS/VS* is found in individuals older than 88 years [30], speaking against the VS being a simple longevity allele. In fact, one case report of a de novo gene translocation causing what was interpreted as but not yet proven to be high plasma Klotho activity, is associated with a syndrome of hypophosphatemic rickets, hyperparathyroidism, and skeletal abnormalities [31]. The pathophysiology of this presumed high Klotho state is not known.

One ELISA kit recently developed to determine plasma Klotho, found levels that are two-fold higher in normal children than adults [32]. Likewise, Semba et al. [33] used the same kit to determine plasma Klotho and found an association of low plasma Klotho with higher overall mortality in a community-based population. Furthermore, the same group found an independent association of higher plasma Klotho with low incidence of cardiovascular disease (table 1), but no relationship between plasma Klotho and renal function (eGFR) [34].

Whether and how blood Klotho is downregulated in elderly humans remains to be examined. One possibility is that loss of nephrons in the aged kidney reduces Klotho expression and shedding into circulation. Another is that renal Klotho reduction is a part of systemic Klotho reduction by an internal 'clock' which renders the kidneys more susceptible to cumulative insults.

CKD Patients

Renal *Klotho* RNA is markedly decreased in CKD patients with etiologies (table 1) including obstructive nephropathy, rejected transplanted kidneys, diabetic nephropathy, chronic glomerulonephritis, and unknown causes [20]. However, data on plasma Klotho levels in human CKD is limited. Using a recently generated ELISA, Yamazaki et al. showed that plasma Klotho correlated with S_{Cr}, BUN and FGF23 [32], suggesting plasma Klotho might be affected by renal function even though that study did not specifically enroll CKD patients. Using the same kit, another group showed 3- to 4-fold higher plasma Klotho in CKD patients compared to healthy volunteers (table 1) [35]. Larger multicenter cohorts with validated assays are required to define plasma Klotho levels in CKD patients.

While clinical data in human plasma Klotho is yet to be acquired, urinary Klotho levels of CKD patients were shown to decrease at a very early stage and sustainably reduced with progression of CKD (table 1) [5]. Klotho levels in plasma, urine and kidney are decreased in parallel in a rodent CKD model (table 1) [5] but this parallel relationship remains to be studied in humans.

Rodent CKD Models

Subtotal (5/6th) nephrectomized rats have low *Klotho* mRNA expression in the kidney 8 weeks after surgery (table 1) [26]. Another CKD rodent model induced by unilateral nephrectomy plus contralateral ischemic reperfusion injury (Npx + IRI) followed by high dietary phosphate intake has also downregulated renal Klotho protein and mRNA expression (table 1) [5]. In addition to the moderate increase in S_{Cr}, hyperphosphatemia, and anemia, ectopic calcification is observed in soft tissues including the kidney, aorta, and heart. Apolipoprotein E-deficient (*apo-E*$^{-/-}$) mice rendered uremic by uninephrectomy and contralateral electrocautery have lower renal Klotho expression compared to *apo-E*$^{-/-}$ non-uremic mice (table 1) [22]. The degree of reduction in Klotho is similar in magnitude in plasma, kidney, and urine [5] indicating that CKD is a sta1te of 'pan-Klotho' deficiency. One implication is that measurement of urinary Klotho might be a simpler and easier surrogate to plasma Klotho.

In addition to CKD due to ablation of renal mass, rodents with immune-mediated chronic glomerulonephritis also have reduced renal *Klotho* mRNA (table 1) [17]. Klotho overexpression improves renal function and ameliorates renal histology, which are associated with less superoxide anion generation and lipid peroxidation, and with decreased levels of cell senescence markers, mitochondrial DNA fragmentation, and apoptosis. Thus, Klotho protein might serve as a renoprotective factor by diminishing oxidative stress, cell senescence and apoptosis [17].

The OLETF rat, a model for the metabolic syndrome, has lower *Klotho* mRNA in the kidneys (table 1) [23, 26]. The administration of the peroxisome proliferator-activated receptor-γ agonist, thiazolidinedione, increases renal *Klotho* mRNA, attenuates abnormal lipid and glucose metabolism, and reduces systolic blood pressure [23]. Whether thiazolidinedione directly stimulates renal Klotho expression on this model is not known but it is conceivable that the increase in Klotho is what preserves kidney function and improves lipid metabolism.

Klotho protein in the kidney is decreased in a streptozotocin-induced diabetic nephropathy model (table 1) [36]. Both insulin and phloridzin correct hyperglycemia, reverse renal Klotho reduction, and improve kidney function and histology. An in vitro study showed that high glucose medium incubation suppresses Klotho expression in canine kidney cells. Insulin is able to shed extracellular domain of Klotho [10] and may increase Klotho release into the blood which may be a novel mechanism of insulin action in alleviation of diabetic ne-

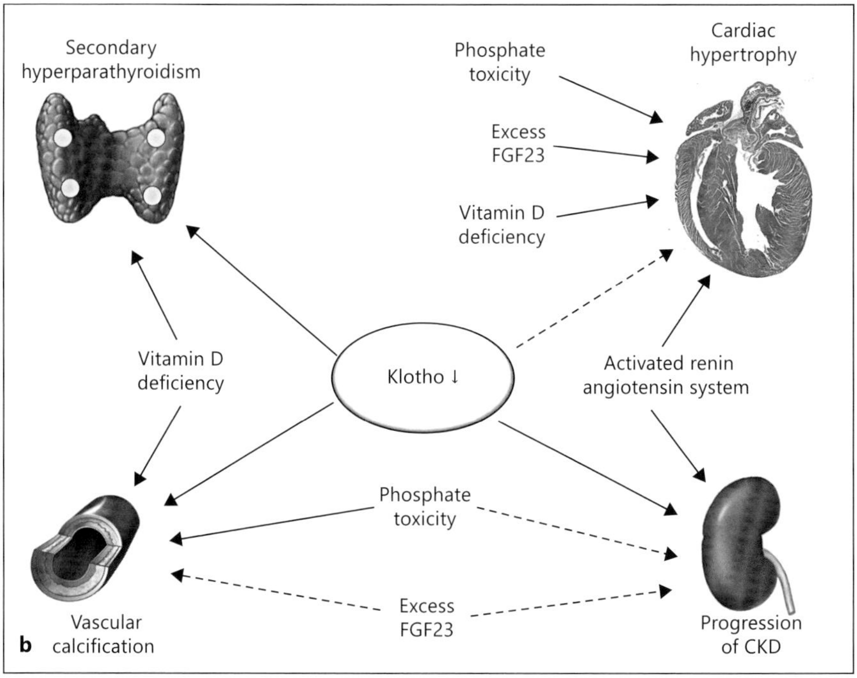

Fig. 1. b Pathogenic role of Klotho deficiency and interaction with other mineral metabolism disturbances in chronic kidney diseases: in the uremic parathyroid gland, FGFR1 and Klotho expression are downregulated which renders parathyroid gland resistant to the suppressive effect of FGF23 which triggers and promotes secondary hyperparathyroidism. The high blood phosphate and FGF23, activated renin-angiotensin system activation, and low blood vitamin D in concert contribute to cardiomyopathy. Whether Klotho plays a direct role in cardiac hypertrophy remains to be clarified (dashed line). High phosphate induces vascular smooth muscle cells reprogramming to osteoblast/chondrocytes, low blood Klotho contributes to failure of suppression of high phosphate-induced calcification. The direct effect of FGF23 on vascular calcification remains to be explored (dashed line). Low blood vitamin D accelerates vascular calcification. Low blood and renal Klotho facilitate renal fibrogenesis, impair tissue regeneration, and consequently promote chronic progression of CKD. Whether and how high FGF23 and phosphate in blood induces renal fibrosis is still unknown (dashed line).

in renal tubular cells of Dahl salt-resistant hypertensive rats [45]. Antioxidant and NF-κB inhibitors reverse the suppressive effect of indoxyl sulfate on Klotho expression in rats and HK2 cells. Indoxyl sulfate may downregulate renal Klotho in CKD (fig. 1a) through stimulation of oxidative pathways and NF-κB [44].

Renin-Angiotensin-Aldosterone (RAA) System

Ang II and aldosterone suppress Klotho expression in the kidney and kidney cell lines [46]. Long-term infusion of Ang II increases blood pressure and downregulates renal Klotho mRNA and protein. Administration of norepinephrine [15] or inhibitor of NO synthesis [6] cause a comparable hypertensive effect, but do not affect renal Klotho expression. Low-dose Ang II does not affect systolic blood pressure, but reduces renal *Klotho* mRNA in rats [15]. In addition, both the angiotensin type I receptor antagonist losartan, and the vasodilator hydralazine block the Ang II-induced rise in blood pressure, but only losartan blocks the Ang II–induced decrease in Klotho. In vivo *Klotho* gene transfer protects the kidney against AII-induced damage (fig. 1a) [15]. Losartan also upregulates the expression of Klotho and improves renal histological alteration in mice of chronic cyclosporine nephropathy [47].

It was proposed that Ang II-induced oxidative stress contributes to downregulation of Klotho expression, because free radical scavenging prevents Klotho depletion by both oxidative stress [9] and Ang II infusion [15]. The interplay between RAA activation, phosphate, and the vitamin D-FGF23 axis may amplify the downregulation of Klotho expression in CKD (fig. 1a). VDR null mice have hypertension and cardiac hypertrophy [48], higher renin expression and plasma Ang II compared to *WT* mice. In *WT* mice, inhibition of 1,25-$(OH)_2VD_3$ synthesis leads to increase in renin expression, whereas 1,25-$(OH)_2VD_3$ injection suppresses renin [49]. Hence, 1,25-$(OH)_2VD_3$ is a negative endocrine regulator of the RAA system (fig. 1a). Active vitamin D_3 upregulates Klotho expression in mouse kidney [18] and in HEK 293 cells [46], and reverses aldosterone-induced *Klotho* mRNA reduction [46]. Therefore, combination of RAA inhibition and supplementation of active vitamin D_3 may provide synergistic effects on upregulation of Klotho expression.

Role of Klotho Deficiency in Progression of Chronic Kidney Disease

Klotho deficiency appears not to be a mere biomarker for CKD but pathogenic for CKD development, progression and complications (fig. 1b). To elaborate on this notion, we will examine the possible ways by which Klotho deficiency can affect the progression of CKD.

Rending Renal Tissues More Vulnerable to Injury

Klotho$^{-/-}$ mice have increased senescence of progenitor cells [14]. Knock-down of endogenous Klotho promotes senescence in primary human fibroblasts [16] and exogenous Klotho decreases senescence in endothelial cells [50] and fibroblasts [16]. Cell senescence induced by Klotho depletion may be mediated by upregulation of Wnt signal activity [14]. Administration of exogenous Wnt could accelerate cell senescence in vivo and in vitro. Soluble Klotho binds to various Wnt family members and inhibits their biological activity [14]. One recent study shows that intracellular Klotho suppresses cell senescence by inhibiting retinoic acid-inducible gene-I-induced expression of IL-6 and IL-8 both in vitro and in vivo [51].

Cell senescence is a complicated process present not only in normal aging but also in pathophysiological states [17, 52]. Klotho deficiency in CKD would enhance cell senescence induced by oxidative stress [17, 52] and accelerated senescence and resultant apoptosis and stem cell depletion might decrease the kidney's ability to defend against renal insults and impair regeneration.

Promote Renal Fibrogenesis

Renal fibrosis is a prominent histological characteristic in CKD, and likely contributes to chronic progression; and TGF-β_1 is a major player in renal fibrosis [53]. The renal fibrosis in unilateral ureteral obstruction (UUO) is accompanied by upregulation of TGF-β_1, and downregulation of Klotho mRNA and protein. These alterations are exaggerated in *Klotho*$^{+/-}$ mice [54]. Klotho alleviates renal fibrosis induced by UUO and suppresses expression of fibrosis markers and TGF-β_1 target genes (Snail, Twist), by inhibiting TGF-β_1 binding to type-II TGF-β receptor [54]. Furthermore, Klotho suppresses TGF-β_1-induced Smad2 phosphorylation in a rat tubular epithelial cell line suggesting that Klotho protein inhibits fibrosis primarily through inhibiting TGF-β_1 signaling [54].

The combination of promotion of renal fibrogenesis and retardation of renal tissue regeneration can potentially promote chronic progression (fig. 1b) and Klotho supplementation can impart beneficial impact in preventing and slowing down CKD progression.

Contributions of Klotho Deficiency to Complications of Chronic Kidney Disease

An emerging hypothesis is that Klotho deficiency not only worsens renal disease but also exacerbates extra-renal complications in CKD.

Ectopic Calcification

Cardiovascular calcification is a major underlying factor for high mortality and morbidity in CKD [55]. Vascular calcification is a dynamic process resulting from the imbalance between promoters and inhibitors, FGF23 and Klotho are novel contributors to ectopic calcification in soft tissues including aorta (fig. 1b) [4, 5, 56, 57].

Klotho$^{-/-}$ mice have extensive ectopic calcification in soft tissue which is also observed in CKD subjects, suggesting a potential pathogenic association between

Klotho deficiency and calcification. Augmentation of Klotho by supplementation or genetic manipulation significantly inhibits vascular calcification in CKD animals [5]. Elevated plasma Pi correlates with vascular calcification in CKD, and multiple pathogenetic roles of high Pi have been proposed to induce or exacerbate vascular calcification: (1) high ambient Pi induces vascular smooth muscle cells reprogramming to a phenotype resembling osteoblast and osteochondrocyte in a type 3 Na^+-coupled transporters (Pit1 and Pit2)-dependent manner [58], and (2) high Pi causes cell apoptosis and death and resultant vesicle release for deposition on the matrix [59]. Klotho was shown to suppress this reprogramming induced by a high Pi in rat cell lines in vitro [5]. Klotho also serves as an anti-inflammatory modulator [24] in vascular endothelial cells and smooth muscle cells [16, 52].

FGF23-deficient mice have ectopic calcification in soft tissues including vasculature, raising the interesting notion that FGF23 may be protective for the vasculature. However, the high blood FGF23 in CKD patients challenges this model. One possible explanation might be that the target organs lose their response to FGF23 in CKD. Klotho and FGFR1/3 expression are downregulated when human aorta derived-smooth muscle cells are incubated with high phosphate and calcium. This concept was reinforced by the recent in vitro finding of expression of the FGF23, FGFR1/3 and Klotho in human aorta [56, 57]. Vascular smooth muscle cells undergoing osteogenic/chondrogenic trans-differentiation under the effect of uremic milieu including high Pi lose the ability to respond to FGF23 (fig. 1b) [56]. Another possibility is that extremely high FGF23 may exert pathological action via activation of off-target genes to induce vascular calcification in the setting of low blood Klotho.

Uremic Cardiomyopathy

Cardiac hypertrophy, especially left ventricular hypertrophy, is frequently encountered in CKD patients. In addition to traditional risk factors, novel factors including RAA system activation [60], vitamin D deficiency [61], high FGF23 [62], and low Klotho [62] are implicated in the cardiomyopathy of CKD (fig. 1b).

In the heart, Klotho is expressed solely at the sinoatrial node [3]. The high rate of sudden death of *Klotho*$^{-/-}$ mice under stress may partly be caused by dysrhythmia from sinoatrial node dysfunction [3]. However, no degenerative structural change was found in sinoatrial node of *Klotho*$^{-/-}$ mice, suggesting that Klotho may help maintain the functional integrity of the sinoatrial node [3]. Whether and how soluble Klotho functions as regulator of the heart pacemaker is still unclear.

The association of Klotho deficiency and cardiac hypertrophy was recently shown [62]. This study documented that even heterozygous Klotho haploinsuuficiency is associated with left ventricular hypertrophy which is not hypertension-dependent. Elevated cardiac or/and systemic FGF23 is a direct intermediate, because both intramyocardial and intravenous injection of FGF23 result in cardiac hypertrophy in *WT* mice, and in an established animal model of CKD, treatment with an FGF-receptor blocker attenuates cardiac hypertrophy, but does not change blood pressure [62]. These results unveil a Klotho-independent role for FGF23 in the pathogenesis of cardiac hypertrophy. In addition to FGF23, aberrant vitamin D and phosphate metabolism and cardiac RAA activation in Klotho deficiency situation may also participate in cardiac hypertrophy in CKD (fig. 1b).

Secondary Hyperparathyroidism
In addition to aberrant mineral metabolism (Pi and Ca, vitamin D), high FGF23 and low Klotho may contribute to secondary hyperparathyroidism. CKD patients have elevated plasma level and activity of FGF23 [62] and low FGFR1 and Klotho expression in parathyroid gland [63]. However, one study did find increased Klotho protein in uremic parathyroid gland, which may reflect Klotho expression at the very early stage of secondary hyperparathyroidism [64].

In the physiological setting with normal Klotho and FGFR expression, FGF23 decreases PTH production, increases expression of both the parathyroid calcium-sensing receptor and the vitamin D receptor, and suppresses cell proliferation [65]. In contrast, in the uremic setting with low Klotho and FGFR1 expression in parathyroid glands, FGF23 loses its inhibitory actions on the gland, and fails to increase calcium-sensing receptor and vitamin D receptor [65]. Moreover, FGF23 administration fails to reduce blood PTH levels in CKD animals [66], indicating parathyroid resistance to FGF23, which may result from low expression of FGFR1 and Klotho in uremic hyperplastic parathyroid glands (fig. 1b) [65].

Conclusion and Perspectives

With the recent escalation of research of Klotho on renal physiology and pathophysiology, the understanding of the relationship of Klotho with kidney disease has been advanced considerably. Animal data and clinical observation to date is overwhelmingly strong to indicate that Klotho is not merely an early biomarker for CKD, but also a pathogenetic intermediate for CKD progression and complications development. Animal studies unequivocally showed that Klotho supplementation may provide novel therapy for acute kidney injury patients to retard or block its progression to CKD and for CKD by slowing progression as well as preventing and reversing extra-renal complications.

The database on plasma Klotho levels in human CKD calls for robust, validated, standardized, and high-throughput assay(s) suitable for large-scale population-based cross-sectional and longitudinal studies. The publications thus

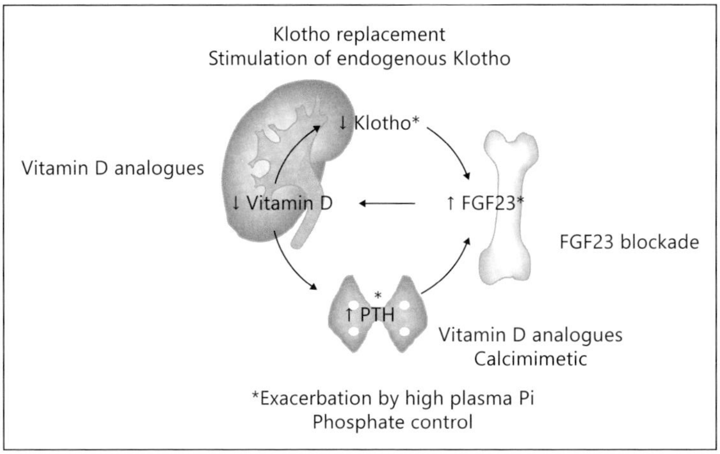

Fig. 2. Proposed therapeutic strategies in CKD. Five principal mineral disturbances in CKD: Klotho deficiency, excess FGF23, secondary hyperparathyroidism, vitamin D deficiency, and phosphate retention are shown as interacting and self-amplifying setting off a vortex of downhill spiral. Shown in italics are means to interrupt these pathways at specific points. Each maneuver has the potential of breaking the spiral. In a given patient, a particular combination of prescriptions can be tailored to best suit that individual. The prescription will also not be static in the sense that it has to be constantly readjusted with time as the patient's pathophysiology evolves and responds to therapy.

far, while informative, are from cross-sectional study and from single center with small sample size and to date use assays that cannot yet be standardized. Plasma or urinary Klotho should be explored as markers for presence of renal disease, predictors for progression, and extra-renal complications, particularly soft tissue calcification.

Beside the important observational database, it is critical to know how to translate the preclinical therapeutic results of Klotho in animal experiments to human CKD. Identification of indications for Klotho treatment will be challenging. When should we start Klotho treatment in CKD patients? As low renal Klotho and urinary Klotho may be present in very earlier stage of CKD, it is reasonable to start Klotho administration as early as possible. If that is the case, what are the therapeutic targets and how does one monitor therapy, since the effects of Klotho is so pleiotropic. Can Klotho replacement be still beneficial in advanced CKD or ERSD? Perhaps the extrarenal effects of Klotho still warrant its replacement even if there are no renal effects or salvage.

The most important concept is that we will likely individualize therapy in mineral disturbances in CKD. Figure 2 depicts a model where the principal mineral disturbances in CKD amplify each other thus creating a self-intensifying vortex that spirals downhill accelerating renal demise. In a given patient at a particular stage, one can envision that interruption of one or more factors may take precedence over others. Klotho supplementation therapy will most likely be part of a more complex regimen that is tailored to a particular patient.

It is likely that Klotho downregulation in CKD/ERSD is multifactorial (fig. 1) and each patient has different etiology of kidney disease and complication, and may have specific factor(s) that downregulate Klotho expression. Thus in addition to direct supplementation of Klotho protein, control or removal of those negative factors may synergistically increase Klotho efficacy or decrease Klotho dosage. Thus individualized regimens may include Klotho, renin-angiotensin system inhibitors, HMG CoA reductase inhibitors, vitamin D analogues, peroxisome proliferator-activated receptor-γ agonists, or antioxidants to enhance endogenous Klotho.

References

1 Coresh J, Selvin E, Stevens LA, Manzi J, Kusek JW, Eggers P, Van Lente F, Levey AS: Prevalence of chronic kidney disease in the United States. JAMA 2007;298:2038–2047.

2 Levey AS, Atkins R, Coresh J, Cohen EP, Collins AJ, Eckardt KU, Nahas ME, Jaber BL, Jadoul M, Levin A, Powe NR, Rossert J, Wheeler DC, Lameire N, Eknoyan G: Chronic kidney disease as a global public health problem: approaches and initiatives: a position statement from Kidney Disease Improving Global Outcomes. Kidney Int 2007;72:247–259.

3 Takeshita K, Fujimori T, Kurotaki Y, Honjo H, Tsujikawa H, Yasui K, Lee JK, Kamiya K, Kitaichi K, Yamamoto K, Ito M, Kondo T, Iino S, Inden Y, Hirai M, Murohara T, Kodama I, Nabeshima Y: Sinoatrial node dysfunction and early unexpected death of mice with a defect of klotho gene expression. Circulation 2004;109:1776–1782.

4 Kuro-o M, Matsumura Y, Aizawa H, Kawaguchi H, Suga T, Utsugi T, Ohyama Y, Kurabayashi M, Kaname T, Kume E, Iwasaki H, Iida A, Shiraki-Iida T, Nishikawa S, Nagai R, Nabeshima YI: Mutation of the mouse klotho gene leads to a syndrome resembling ageing. Nature 1997;390:45–51.

5 Hu MC, Shi M, Zhang J, Quinones H, Griffith C, Kuro-o M, Moe OW: Klotho deficiency causes vascular calcification in chronic kidney disease. J Am Soc Nephrol 2011;22:124–136.

6 Ohyama Y, Kurabayashi M, Masuda H, Nakamura T, Aihara Y, Kaname T, Suga T, Arai M, Aizawa H, Matsumura Y, Kuro-o M, Nabeshima Y, Nagail R: Molecular cloning of rat klotho cDNA: markedly decreased expression of klotho by acute inflammatory stress. Biochem Biophys Res Commun 1998;251:920–925.

7 Forster RE, Jurutka PW, Hsieh JC, Haussler CA, Lowmiller CL, Kaneko I, Haussler MR, Kerr Whitfield G: Vitamin D receptor controls expression of the anti-aging klotho gene in mouse and human renal cells. Biochem Biophys Res Commun 2011;414:557–562.

8 Hu MC, Shi M, Zhang J, Pastor J, Nakatani T, Lanske B, Razzaque MS, Rosenblatt KP, Baum MG, Kuro-o M, Moe OW: Klotho: a novel phosphaturic substance acting as an autocrine enzyme in the renal proximal tubule. FASEB J 2010;24:3438–3450.

9 Mitobe M, Yoshida T, Sugiura H, Shirota S, Tsuchiya K, Nihei H: Oxidative stress decreases klotho expression in a mouse kidney cell line. Nephron Exp Nephrol 2005;101:e67–e74.

10 Chen CD, Podvin S, Gillespie E, Leeman SE, Abraham CR: Insulin stimulates the cleavage and release of the extracellular domain of Klotho by ADAM10 and ADAM17. Proc Natl Acad Sci USA 2007;104:19796–19801.

11 Imura A, Iwano A, Tohyama O, Tsuji Y, Nozaki K, Hashimoto N, Fujimori T, Nabeshima Y: Secreted Klotho protein in sera and CSF: implication for post-translational cleavage in release of Klotho protein from cell membrane. FEBS Lett 2004;565:143–147.

12 Urakawa I, Yamazaki Y, Shimada T, Iijima K, Hasegawa H, Okawa K, Fujita T, Fukumoto S, Yamashita T: Klotho converts canonical FGF receptor into a specific receptor for FGF23. Nature 2006;444:770–774.
13 Chang Q, Hoefs S, van der Kemp AW, Topala CN, Bindels RJ, Hoenderop JG: The beta-glucuronidase klotho hydrolyzes and activates the TRPV5 channel. Science 2005;310:490–493.
14 Liu H, Fergusson MM, Castilho RM, Liu J, Cao L, Chen J, Malide D, Rovira II, Schimel D, Kuo CJ, Gutkind JS, Hwang PM, Finkel T: Augmented Wnt signaling in a mammalian model of accelerated aging. Science 2007;317: 803–806.
15 Mitani H, Ishizaka N, Aizawa T, Ohno M, Usui S, Suzuki T, Amaki T, Mori I, Nakamura Y, Sato M, Nangaku M, Hirata Y, Nagai R: In vivo klotho gene transfer ameliorates angiotensin II-induced renal damage. Hypertension 2002;39:838–843.
16 de Oliveira RM: Klotho RNAi induces premature senescence of human cells via a p53/ p21 dependent pathway. FEBS Lett 2006;580: 5753–5758.
17 Haruna Y, Kashihara N, Satoh M, Tomita N, Namikoshi T, Sasaki T, Fujimori T, Xie P, Kanwar YS: Amelioration of progressive renal injury by genetic manipulation of Klotho gene. Proc Natl Acad Sci USA 2007;104:2331–2336.
18 Tsujikawa H, Kurotaki Y, Fujimori T, Fukuda K, Nabeshima Y: Klotho, a gene related to a syndrome resembling human premature aging, functions in a negative regulatory circuit of vitamin D endocrine system. Mol Endocrinol 2003;17:2393–2403.
19 Silver J, Naveh-Many T: FGF23 and the parathyroid glands. Pediatr Nephrol 2010;25: 2241–2245.
20 Koh N, Fujimori T, Nishiguchi S, Tamori A, Shiomi S, Nakatani T, Sugimura K, Kishimoto T, Kinoshita S, Kuroki T, Nabeshima Y: Severely reduced production of klotho in human chronic renal failure kidney. Biochem Biophys Res Commun 2001;280:1015–1020.
21 Hu MC, Kuro-o M, Moe OW: Klotho and kidney disease. J Nephrol 2010;23(suppl 16): S136–S144.
22 Yu J, Deng M, Zhao J, Huang L: Decreased expression of klotho gene in uremic atherosclerosis in apolipoprotein E-deficient mice. Biochem Biophys Res Commun 2010;391: 261–266.
23 Yamagishi T, Saito Y, Nakamura T, Takeda S, Kanai H, Sumino H, Kuro-o M, Nabeshima Y, Kurabayashi M, Nagai R: Troglitazone improves endothelial function and augments renal klotho mRNA expression in Otsuka Long-Evans Tokushima Fatty (OLETF) rats with multiple atherogenic risk factors. Hypertens Res 2001;24:705–709.
24 Zhao Y, Banerjee S, Dey N, LeJeune WS, Sarkar PS, Brobey R, Rosenblatt KP, Tilton RG, Choudhary S: Klotho depletion contributes to increased inflammation in kidney of the *db/db* mouse model of diabetes via RelA (serine)536 phosphorylation. Diabetes 2011; 60:1907–1916.
25 Wang Y, Sun Z: Klotho gene delivery prevents the progression of spontaneous hypertension and renal damage. Hypertension 2009;54:810–817.
26 Aizawa H, Saito Y, Nakamura T, Inoue M, Imanari T, Ohyama Y, Matsumura Y, Masuda H, Oba S, Mise N, Kimura K, Hasegawa A, Kurabayashi M, Kuro-o M, Nabeshima Y, Nagai R: Downregulation of the Klotho gene in the kidney under sustained circulatory stress in rats. Biochem Biophys Res Commun 1998;249:865–871.
27 Zuo Z, Lei H, Wang X, Wang Y, Sonntag W, Sun Z: Aging-related kidney damage is associated with a decrease in klotho expression and an increase in superoxide production. Age (Dordr) 2011;33:261–274.
28 Manya H, Akasaka-Manya K, Endo T: Klotho protein deficiency and aging. Geriatr Gerontol Int 2010;10(suppl 1):S80–S87.
29 Arking DE, Krebsova A, Macek M Sr, Macek M Jr, Arking A, Mian IS, Fried L, Hamosh A, Dey S, McIntosh I, Dietz HC: Association of human aging with a functional variant of klotho. Proc Natl Acad Sci USA 2002;99:856–861.
30 Invidia L, Salvioli S, Altilia S, Pierini M, Panourgia MP, Monti D, De Rango F, Passarino G, Franceschi C: The frequency of Klotho KL-VS polymorphism in a large Italian population, from young subjects to centenarians, suggests the presence of specific time windows for its effect. Biogerontology 2010;11:67–73.
31 Brownstein CA, Adler F, Nelson-Williams C, Iijima J, Li P, Imura A, Nabeshima Y, Reyes-Mugica M, Carpenter TO, Lifton RP: A translocation causing increased alpha-klotho level results in hypophosphatemic rickets and hyperparathyroidism. Proc Natl Acad Sci USA 2008;105:3455–3460.

Razzaque MS (ed): Phosphate and Vitamin D in Chronic Kidney Disease.
Contrib Nephrol. Basel, Karger, 2013, vol 180, pp 64–73 (DOI: 10.1159/000346782)

Nuclear Receptor LXR: A New Partner for Sodium-Dependent Phosphate Cotransporters

Yupanqui Caldas[a, b] · Hector Giral[a] · Victor Sorribas[b] · Moshe Levi[a]

[a]University of Colorado Denver, Aurora, Colo., USA; [b]University of Zaragoza, Zaragoza, Spain

Abstract

New pharmaceutical research approaches are focusing on trying to alleviate the perturbed phosphate (Pi) homeostasis associated with the onset of chronic kidney disease; this includes activation of some of the nuclear receptors. We have recently reported the down regulation of the intestinal and renal sodium-phosphate (NaPi) cotransporters by the liver X receptor (LXR) agonists, and the consequent decrease of the serum Pi levels. In this review, we describe our current knowledge of the different proteins involved in the renal and intestinal actions of LXR.

Phosphorous is essential for life due to its role in many biological process [1]. Phosphorus is primarily found as inorganic phosphate (Pi) and stored as hydroxyapatite in the bone, where 80–85% of the total phosphate is found. Around 0.1% Pi is present in the extracellular fluids, where several mechanisms are required to maintain a concentration of free Pi at about 1.1 mM [2, 3].

Pi homeostasis is maintained through the activity of synchronized mechanisms involving intestinal absorption, retention-release from the bone, and excretion-reabsorption at the kidney. Changes in the intestinal Pi absorption rate and renal reabsorption rate are achieved by modifying the abundance of NaPi cotransporters in these epithelia, the affinity of the transporters for the substrate, the stability of the transporters in the membrane (e.g. interactions with the PDZ domain proteins) and the cell membrane lipid composition [2], which could influence the preferred conformational state of the transporter, thus modulating its activity and interactions with regulatory cofactors [3].

Most of the present knowledge about the control of Pi homeostasis has been restricted to the role of the kidney. This organ is considered to be the main controller of phosphate homeostasis, in which several regulatory factors such as dietary phosphate, hormones such as fibroblast growth factor 23 (FGF23) and primarily (PTH), fibroblast growth factor (FGF)23, and VDR (a nuclear receptor, NR) alter the abundance of the NaPi transporters in the proximal tubular cells [1, 3]. Additionally, the use of a proximal tubule cell model (Opossum kidney, OK cells) has allowed the study the mechanisms of the renal Pi transport regulation in the kidney. Recently, however, the role that the small intestine plays in Pi homeostasis has become a new focus of research, including both the mechanisms of intestinal Pi absorption and the regulatory mechanism that govern the intestinal Pi transporters. A new knowledge in intestinal Pi absorption physiology will lead to the establishment of new strategies designed to alleviate the hyperphosphatemia associated with chronic and end stage renal disease, through inhibition of the intestinal NaPi transport activity.

The clinical incidence of hyperphosphatemia in the progression of CKD, VC and aging has been explored for several decades. Now, it is becoming more clear that the main therapeutic strategy that alleviates the dysregulated phosphate homeostasis corresponds to the inhibition of the intestinal phosphate absorption, mainly due to the advanced functional decline of the kidney. In this regard, new pharmaceutical approaches are required to reduce the intestinal absorption of inorganic phosphate from the diet.

We have recently reported that activation of the NR liver X receptor (LXR) by two different ligands in kidney and small intestine leads to the decreased expression of the renal and intestinal NaPi cotransporters, and, consequently, to both, an increase in the urinary Pi excretion, and a small decrease in the serum Pi concentration [4] (fig. 1). This was associated with an increased concentration of circulating FGF23, with no changes in plasma PTH or Ca levels; however, the mechanism is still unknown. Despite of that the purpose of this review will be to describe the different factors that are involved in such LXR effects, and the possible biological components that might explain the regulation of the NaPi cotransporters by the LXR.

Renal NaPi Transporters

Kidney is the major regulatory organ for Pi homeostasis, and this is accomplished by adjusting the reabsorption rate of Pi from the primary urine, through the control of the NaPi cotransporter abundance at the brush-border membrane (BBM) of the proximal tubular epithelial cells. Pi transporters from

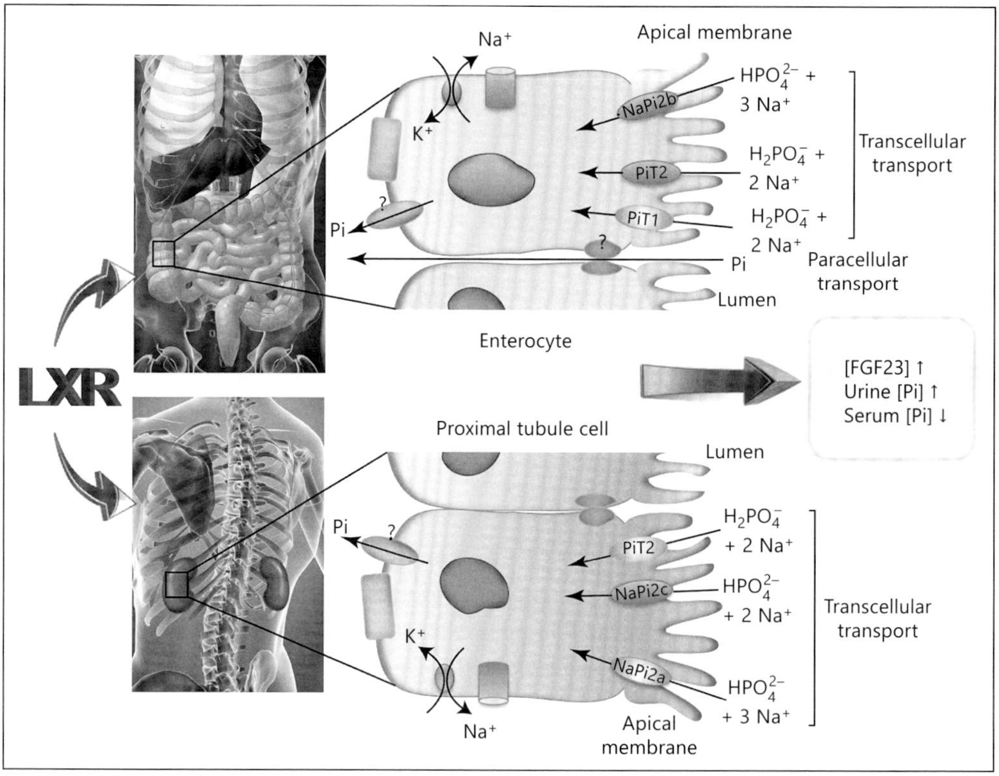

Fig. 1. Activation of LXR in mouse small intestine and kidney has shown to reduce the expression of the intestinal Pi transporter NaPi2b and the renal NaPi2a and NaPi2c. This changes are associated with a increase of FGF23 serum levels, an increase of urinary Pi excretion, and a decrease of serum Pi concentration. This might indicate the pharmacological relevance that LXR could have in the control of hyperphosphatemia in humans.

the three different families of Solute Carries (SLC17, SLC20 and SLC34) have been identified and located in the kidney. The type II subfamily of NaPi cotransporters NaPi2a and NaPi2c have a high affinity for divalent Pi (HPO_4^{2-}).

NaPi2a (SLC34A1) is the main Pi transporter in the kidney, accounting for the majority of the renal Pi handling. NaPi2a was initially cloned by the group of Murer, while NaPi2c (SLC34A3) was latter reported by Segawa et al. [5]. NaPi2a has been the most studied NaPi cotransporter. Basically what we know about NaPi2a is: (1) In rodents, it is responsible for 70% of Pi reabsorption in the kidney, based on studies of the NaPi2a KO mice (Npt2a–/–) [6], but a more recent study has reported a higher percentage [7]; in human, its importance is exemplified by a mutation in the hNpt2a gene that induces renal Fanconi's syndrome and hypophosphatemic rickets [8]. (2) NaPi2a is regulated by changes in the Pi concentrations; being upregulated in the BBM as a response to low phosphate concentrations in serum [9]. (3) Some PDZ domain-containing proteins, such as the NHERF family are required for NaPi2a trafficking and/or stabilization in the apical membrane. (4) NaPi2a responds to many hormonal stimuli, primarily to parathyroid hormone (PTH) which targets NaPi2a for internalization through activation of a cAMP-PKA/PKC-ERK dependent mechanism, and this process occur much faster than internalization of NaPi2c after PTH treatment [10]. (5) Phosphatonins, primarily FGF23, regulate the expression of NaPi2a. NaPi2c is a more recent discovered NaPi cotransporter that was initially shown to be related to growth process in young animals, since it is highly expressed in rodents during unweaning in contrast to NaPi2a which shows a more age-independent expression [5].

NaPi2c also responds to the same regulatory factors as NaPi2a [11]; however, the response to PTH and serum Pi concentration seem to follow different mechanisms and different time frames, in which the response of NaPi2a to hormonal and nonhormonal stimulus seems to be faster than for NaPi2c [12]. In this respect, we have hypothesized that NaPi2a and NaPi2c interact differently with NHERF1 and PDZK1 respectably, PDZ proteins that are important for the regulation of these NaPi co-transporters in the BBM [13].

Finally, the SLC20 family of Pi transporters, PiT1 (SLC20A1) and PiT2 (SLC20A2), were identified over 20 years ago as virus receptors, and later they were characterized as a Pi transporters [14]. Both transporters are expressed in the kidney but only PiT2 has been clearly shown to be expressed in the BBM of proximal tubular cells [15] and also to respond to changes in dietary Pi. Along with the type II NaPi transporters, PiT2 is upregulated by a low Pi diet. However, in contrast to the type II NaPi transporters, the type III transporters have a preference for monobasic phosphate ($H_2PO_4^-$) and therefore show higher Pi transport at low pH, making the function of the type III transporters distinct from the type II NaPi transporters [7].

Intestinal NaPi Transporters

For decades it was believed that the intestine was a passive organ that allows Pi uptake through a nonsaturable paracellular transport, and that Na-dependent transport had a minimal role in total intestinal Pi absorption [16]. Since then, many other studies have confirmed the presence of an Na-dependent component (Pi transport) of intestinal absorption. In fact, NaPi2b (SLC34A2), a type II Na-Pi cotransporter, is presently considered the transporter responsible of the main Na-dependent Pi absorption in the small intestine [17]. However, despite the fact that the expression of both type III PiT1 and PiT2 have been reported in the small intestine of rodents, their role in total Pi absorption and their regulation, are still unclear.

Similar to the other type II NaPi transporters, NaPi2b presents a preference for the dibasic Pi solute which carries along with 3 sodium ions for an electrogenic transport cycle, with a relative high affinity for Pi (K_m of approximately 10 μM). This fact has at times been misinterpreted, when assuming that a high affinity implies a necessary early saturation of transport. However, saturation is only determined by the capacity (i.e. V_{max}) of the transport system, and michaelian capacity is only determined by the number of transporters expressed in the BBM. Therefore, the astonishing high affinity of NaPi2b should not question the relevance of the transporter in the intestinal Pi absorption but, on the contrary, we should try to understand the reason of having such high affinity, when the intestinal content of Pi is usually in the high millimolar range [18]. One explanation is the effect of pH, as NaPi2b transports divalent Pi. As a polyprotic acid, Pi has several dissociation constants, and according to the Henderson-Hasselbach equation, at the physiological jejunal pH, near 6.0 [19], only 6% of Pi is divalent, while 94% is monovalent, i.e. a substrate for type III NaPi transporters. The high affinity of NaPi2b is therefore explained by the small percentage of divalent Pi at such pH. Also, as dissociation is a dynamic process, as soon as divalent Pi is absorbed more monovalent Pi is converted into divalent to maintain Pi species equilibrium, therefore providing new substrate for NaPi2b.

In mouse, NaPi2b mainly expressed in the distal section on the small intestine (ileum, second half) which contrast to the expression pattern in rat and humans: In these species, NaPi2b is most abundant in the jejunum and duodenum (first half), but this is not the only difference among these species [20]. Mutation of NaPi2b in humans has minimum effects on Pi homeostasis. In contrast, a total NaPi2b KO mouse presents an embryonic lethal phenotype. An inducible NaPi2bFlox-Cre KO mouse adapted to an acute high Pi diet exhibits a lower Pi intestinal absorption (around 50%) with correspondingly low levels of serum Pi. This is also observed in WT animals treated with nicotinamide, an agent also known to reduce serum Pi in patients and to decrease expression NaPi2b. Additionally, the inducible NaPi2b KO mouse shows a radical reduction of Na-dependent Pi uptake in ileum in vitro [21, 22]. All these findings provide evidence on the role of NaPi2b in the intestinal Pi absorption.

Similar to renal NaPi transporters, aging and dietary Pi intake also regulate NaPi2b in the intestine. Nevertheless, dietary Pi and vitamin D are the most important regulatory factors of intestinal (NaPi2b-mediated) Pi absorption. Several groups have shown an upregulation of NaPi2b protein in rodents after administration of a low Pi diet, either chronic or acutely, but these changes seem to be also caused by different regulatory mechanisms in the different segments of the small intestine [17, 20]. NaPi2b is also regulated by vitamin D, and initial reports suggested that the increased intestinal Pi absorption (and NaPi2b expression) under low Pi diets depended on the higher synthesis of vitamin D through upregulation of the renal 1alpha-OHase [20]. Later studies in VDR null mice and 1alpha-OHase null mice revealed, however, that intestinal Pi adaptation to low Pi diet is still present in both of these mice models Recent studies by Marks et al., in which they compared the expression of NaPi2b in the three different segments of the rat and mouse small intestine after vitamin D treatment, showed an increase of Pi uptake associated with upregulation of the NaPi2b expression exclusively in the jejunum of both rodents, providing evidence for a direct effect of vitamin D on NaPi2b expression [20].

PDZ domain-containing proteins play an important role in the regulation of the renal NaPi transporters. The case of NaPi2b is less clear because until now, the regulation of intestinal Pi transport has not been linked to PDZ interaction. However our recently published data shows that NHERF1 is involved in the regulation of NaPi2b, in a new established intestinal cell line model [23].

Phosphatonins

A new group of hormones termed phosphatonins were initially described in studies performed on patients with tumor-induced osteomalacia [3]. Later, several phosphatonins were described, including secreted frizzled-related protein 4, matrix extracellular phosphoglycoprotein, FGF7, and FGF23, the latter being the best characterized phosphatonin [1, 3, 6]. Little is still known about the molecular mechanisms of phosphatonins on Pi homeostasis, but all they exhibit a strong phosphaturic effect, mainly through the downregulation of the renal NaPi transporters. In the case of the intestinal NaPi transporters, no direct effects have been described on NaPi2b.

FGF23 was discovered in patients suffering from autosomal-dominant hypophosphatemic rickets which presents a mutation on the adhr gene [3]. FGF23 is produced by the osteocytes as a response to high Pi intake, hyperphosphatemia, or high levels of vitamin D [1]. Although the mechanisms behind the action of FGF23 are not as well established as for PTH, it is known that the phosphaturic effect is associated to a decrease of NaPi2a and NaPi2c in the BBM of the proximal tubule cells. This effect is mediated by the FGF receptor 1 (FGFR1) and a coreceptor, Klotho, which activates a signal transduction pathway that involves phosphorylation of ERK1/2. FGF23 also induces the repression of the vitamin D synthesis by activating the 24-hydroxylase and inhibiting the expression of 1alpha-OHase in kidney, thus affecting the intestinal NaPi2b expression; however, a direct effect of FGF23 on NaPi2b has not been reported. This finding also provides additional evidence for the marked differences in the regulatory mechanism between the renal and the intestinal Pi transporters.

A recent report from Berndt et al. [24] has suggested the presence of a new no-identified phosphatonin-like factor present in the small intestine of rats. This factor could play and important role not only through an acute response to high dietary Pi intake (i.e. inducing a decreased renal Pi reabsorption) without affecting the levels of FGF23, PTH, or secreted frizzled-related protein 4, but also working as a sensor, releasing a substance that induce those effects. This was confirmed with further experiments where homogenates of rat duodenal mucosa were collected after high Pi administration, and then infused into the small intestine of normal rats, inducing an increase of urinary Pi excretion rate within minutes. These studies most likely imply a new phosphatonin release from the rat duodenum with quick phosphaturic effects.

Role of Nuclear Receptors in Pi Homeostasis

NR are involved in different biological processes, from cell development, to metabolism and immunology. They are ligand activated transcription factors capable of activating and repressing gene expression through many different mechanisms [25]. Several NR form heterodimer with the retinoid X receptor (RXR) for the biological function, but RXR is also able to form homodimers for its biological process. Other than VDR, several of these type II NR have been related to Pi regulation: pregnane X receptor (PXR, a xenobiotic NR), RAR (vitamin A receptor), and more recently liver X receptor (LXR, cholesterol sensor). From them, VDR is recognized as the most important in Pi homeostasis control.

PXR is a highly promiscuous drug response transcription factor that mainly induces the expression of enzymes and transporters involved in metabolic inactivation and drug excretion, as part of a cellular defense mechanism against xenobiotic toxicity [26]. However, long-term treatment of the antimicrobial drug rifampicin, which actives PXR has shown a cross-activation of VDR target genes as well. With the aim to determine this cross-talking between PXR and VDR, investigators generated a PXR KO model that manifest low Pi plasma concentration and low bone mineral density, with no significant changes in Ca, vitamin D, or FGF23 levels, therefore linking PXR to Pi metabolism. These results are associated with low protein expression of NaPi2b, downregulation of the NaPi2b mRNA levels in isolated intestinal BBM of PXR–/–, compared to WT mice, and no effect on the CaT1 and CaT2 calcium transporters. Further in vitro experiments showed that PXR activation through its agonist rifampicin is able to activate both human and mouse NaPi2b promoter gene in a luciferase assay.

For decades the role of vitamin A (through its metabolite all-trans-retinoic acid, ATRA), in bone mineralization and Ca-Pi metabolism has been known;

however, the direct relationship of the nuclear receptors RAR or RXR with regulation of renal NaPi transporters was only recently elucidated. Toledo et al. [27] initially determined that OK cells treated with ATRA present an increase in Pi uptake. Later, rats fed vitamin A-deficient diets exhibited phosphaturia with no changes in serum Pi, vitamin D, FGF23 or PTH levels. These results were correlated with lower expression of the renal NaPi transporters NaPi2a and NaPi2c in the BBM, and less mRNA of their respective genes. In vitro studies determined that both gene promoters present a consensus direct repeat (DR-5) retinoic acid response element, to which RXR-RAR heterodimer binds, and this was confirmed by electrophoretic mobility shift assay. The NaPi2c promoter, however, presents an additional DR-6 that overlaps half-site sequence with a consensus response element for VDR, providing the possibility of the transcriptional regulation of NaPi2c by either independent or additive actions. As proof of this, luciferase assays in OK cells transfected with the NaPi2c promoter presents activation with vitamin D, or ATRA, and a higher activation rate when using both activating ligands. The authors hypothesized that this higher level of activation might be important during growth and development [28].

The upregulatory properties of these NRs are mediated through a specific binding sequence in the promoter of their target genes. However, the inhibitory effects are less clear, and other mechanisms should to be considered to understand this physiological process. That is the case of LXR and the inhibitory effects on the NaPi transporters. LXR has been mainly known as a transcription factor that regulates reverse cholesterol transport. This not only occurs through the ability of inducing the expression of the ATP-binding cassette transporters (ABC cholesterol transporters), but LXR also activates ApoE and ApoC, and CYP7A1, the enzyme responsible for the transformation of cholesterol into bile acids. LXR is also known for down-regulating the expression of the pro-inflammatory cytokines IL-1β, IL-6, TNFα, and iNOS, mainly through the repression of the NF-κB gene. This downregulation is achieved through a whole set of novel mechanisms, that share with other lipid metabolism-related NRs, the PPARs [25]. The duality of these ligand activated NRs to control both, lipid homeostasis and inflammatory processes make them very attractive possibilities to medical and pharmaceutical research. These NRs mediate their anti-inflammatory cross-talk, referred to as trans-repression, through a mechanism that involves sumoylation, i.e. a posttranslational modification caused by the family of small ubiquitin-related modifiers required for entering the transrepression pathway [29].

Increase in FGF23 expression after LXR's agonist treatment might be the first explanation for the downregulation of the renal NaPi transporters and the consequent increase of the Pi excretion rate. However, LXR agonists also directly inhibit NaPi cotransport activity and apical membrane NaPi protein abundance in OK cells in culture. In addition, preliminary data from our laboratory suggest that activation of the LXR in kidney induces frank changes in the cholesterol concentration in the BBM and this would induce (a) increased membrane fluidity as shown by Laurdan GP measurements, and (b) changes in the interaction of the renal NaPi cotransporters with some of the PDZ domain proteins [unpubl. data]. This is not surprising because we have shown that changes in the membrane lipid composition can alter protein activity as well as protein trafficking [30].

Although changes in cholesterol levels induced by LXR activation can downregulate the intestinal NaPi transport activity and NaPi2b protein dynamics, we cannot discard other putative mechanisms that could be working synergistically, through the transrepression mechanism describe before. Similar mechanisms could be responsible for the transcriptional regulation of NaPi2b gene after LXR activation. However, since this transrepression mechanism requires the sumoylation of LXRs in the C-terminal activation domain for its recruitment to a specific NR corepressor complex, most likely this should be different to the mechanism involved in the regulation of LXR on NaPi2b. This conclusion is based on our recent preliminary results using the VP16-LXRα mouse. These mice show inhibition of intestinal NaPi transport activity and reduced NaPi2b protein. The VP16-LXRα protein expressed in these animals lacks the C-terminal activation domain of LXRα, and therefore the required sumoylation sites for recruitment to the NR corepressor complex. Nevertheless, we cannot completely discard that a different transrepression pathway involving different corepressors and requirements could be partially modulating the observed effects on NaPi activity and NaPi-2b cotransporter.

References

1 Prie D, Urena Torres P, Friedlander G: Latest findings in phosphate homeostasis. Kidney Int 2009;75:882–889.

2 Murer H, Forster I, Biber J: The sodium phosphate cotransporter family SLC34. Pflugers Arch 2004;447:763–767.

3 Biber J, et al: Regulation of phosphate transport in proximal tubules. Pflugers Arch 2009; 458:39–52.

4 Caldas YA, et al: Liver X receptor-activating ligands modulate renal and intestinal sodium-phosphate transporters. Kidney Int 2011; 80:535–544.

5 Segawa H, et al: Growth-related renal type II Na/Pi cotransporter. J Biol Chem 2002;277: 19665–19672.

6 Beck L, et al: Targeted inactivation of Npt2 in mice leads to severe renal phosphate wasting, hypercalciuria, and skeletal abnormalities. Proc Natl Acad Sci USA 1998;95:5372–5377.

7 Villa-Bellosta R, Sorribas V: Compensatory regulation of the sodium/phosphate cotransporters NaPi-IIc (SCL34A3) and Pit-2 (SLC20A2) during Pi deprivation and acidosis. Pflugers Arch 2010;459:499–508.

8 Magen D, et al: A loss-of-function mutation in NaPi-IIa and renal Fanconi's syndrome. N Engl J Med 2010;362:1102–1109.
9 Levi M, et al: Cellular mechanisms of acute and chronic adaptation of rat renal P(i) transporter to alterations in dietary P(i). Am J Physiol 1994;267:F900–F908.
10 Lanzano L, et al: Differential modulation of the molecular dynamics of the type IIa and IIc sodium phosphate cotransporters by parathyroid hormone. Am J Physiol Cell Physiol 2011;301:C850–C861.
11 Miyamoto K, et al: New aspect of renal phosphate reabsorption: the type IIc sodium-dependent phosphate transporter. Am J Nephrol 2007;27:503–515.
12 Segawa H, et al: Parathyroid hormone-dependent endocytosis of renal type IIc Na-Pi cotransporter. Am J Physiol Renal Physiol 2007;292:F395–F403.
13 Giral H, et al: Role of PDZK1 protein in apical membrane expression of renal sodium-coupled phosphate transporters. J Biol Chem 2011;286:15032–15042.
14 Olah Z, et al: The cellular receptor for gibbon ape leukemia virus is a novel high affinity sodium-dependent phosphate transporter. J Biol Chem 1994;269:25426–25431.
15 Villa-Bellosta R, et al: The Na^+-Pi cotransporter PiT-2 (SLC20A2) is expressed in the apical membrane of rat renal proximal tubules and regulated by dietary Pi. Am J Physiol Renal Physiol 2009;296:F691–F699.
16 Sabbagh Y, et al: Intestinal phosphate transport. Adv Chron Kidney Dis 2011;18:85–90.
17 Giral H, et al: Regulation of rat intestinal Na-dependent phosphate transporters by dietary phosphate. Am J Physiol Renal Physiol 2009; 297:F1466–F1475.
18 Kirchner S, et al: Luminal fructose inhibits rat intestinal sodium-phosphate cotransporter gene expression and phosphate uptake. Am J Clin Nutr 2008;87:1028–1038.
19 McNeil NI, Ling KL, Wager J: Mucosal surface pH of the large intestine of the rat and of normal and inflamed large intestine in man. Gut 1987;28:707–713.
20 Marks J, Debnam ES, Unwin RJ: Phosphate homeostasis and the renal-gastrointestinal axis. Am J Physiol Renal Physiol 2010; 299:F285–F296.
21 Sabbagh Y, et al: Intestinal npt2b plays a major role in phosphate absorption and homeostasis. J Am Soc Nephrol 2009;20:2348–2358.
22 Ohi A, et al: Inorganic phosphate homeostasis in sodium-dependent phosphate cotransporter Npt2b/mice. Am J Physiol Renal Physiol 2011;301:F1105–F1113.
23 Giral H, et al: NHE3 regulatory factor 1 (NHERF1) modulates intestinal sodium-dependent phosphate transporter (NaPi-2b) expression in apical microvilli. J Biol Chem 2012;287:35047–35056.
24 Berndt T, et al: Evidence for a signaling axis by which intestinal phosphate rapidly modulates renal phosphate reabsorption. Proc Natl Acad Sci USA 2007;104:11085–11090.
25 Hong C, Tontonoz P: Coordination of inflammation and metabolism by PPAR and LXR nuclear receptors. Curr Opin Genet Dev 2008;18:461–467.
26 Konno Y, et al: Nuclear xenobiotic receptor PXR-null mouse exhibits hypophosphatemia and represses the Na/Pi-cotransporter SLC34A2. Pharmacogenet Genomics 2010;20: 9–17.
27 de Toledo FG, Beers KW, Dousa TP: Pleiotropic upregulation of Na(+)-dependent cotransporters by retinoic acid in opossum kidney cells. Am J Physiol 1997; 273:F438–F444.
28 Masuda M, et al: Regulation of renal sodium-dependent phosphate co-transporter genes (Npt2a and Npt2c) by all-trans-retinoic acid and its receptors. Biochem J 2010;429:583–592.
29 Ghisletti S, et al: Parallel SUMOylation-dependent pathways mediate gene- and signal-specific transrepression by LXRs and PPAR-gamma. Mol Cell 2007;25:57–70.
30 Inoue M, et al: Partitioning of NaPi cotransporter in cholesterol-, sphingomyelin- and glycosphingolipid-enriched membrane domains modulates NaPi protein diffusion, clustering, and activity. J Biol Chem 2004; 279:49160–49171.

Moshe Levi, MD
Division of Renal Diseases and Hypertension, University of Colorado Denver
Research 2, Room 7002, Box C281
12700 East 19th Avenue, Aurora, CO 80045 (USA)
E-Mail Moshe.Levi@ucdenver.edu

Razzaque MS (ed): Phosphate and Vitamin D in Chronic Kidney Disease.
Contrib Nephrol. Basel, Karger, 2013, vol 180, pp 74–85 (DOI: 10.1159/000346784)

Phosphate Toxicity and Vascular Mineralization

Mohammed S. Razzaque

Department of Oral Medicine, Infection and Immunity, Harvard School of Dental Medicine, Boston, Mass., USA; Department of Pathology, Nagasaki University School of Biomedical Sciences, Nagasaki, Japan

Abstract

Vascular calcification or mineralization is a major complication seen in patients with advanced stages of chronic kidney disease (CKD), and it is associated with markedly increased morbidity and mortality. Most of the CKD-related vascular mineralization is attributable to abnormal mineral ion metabolism. Elevated serum calcium and phosphate levels, along with increased calcium-phosphorus byproduct, and the use of active vitamin D metabolites are thought to be the predisposing factors for developing vascular mineralization in patients with CKD. Recent experimental studies have shown that vascular mineralization can be suppressed by reducing serum phosphate levels, even in the presence of extremely high serum calcium and 1,25-dihydroxyvitamin D levels, indicating that reducing 'phosphate toxicity' should be the important therapeutic priority in CKD patients for minimizing the risk of developing vascular mineralization and the disease progression.

Vascular mineralization used to be considered as a passive process related to nonspecific responses, and with minimum clinical significance. However, recent studies have demonstrated that prevention of vascular mineralization is of great clinical importance, as it can induce lethal complications that include but not limited to thrombosis, arterial rupture, and myocardial infarction. Vascular mineralization may take place, either in intimal layer or in medial layer of the arterial walls with distinct molecular events. Intimal vascular mineralization is usually noted in atherosclerotic lesions of muscular arteries, including coronary arteries and the aorta; the atheromatous process usually starts with focal collections of macrophages filled with lipid droplets that evolved to pre-atheromas, and eventually, with time, to atherosclerotic lesions

[1]. Intimal vascular mineralization superimposed with atherosclerotic lesions is often present in the elderly individuals with or without associated complications, such as hypertension, diabetes, and dyslipidemia [2]. In contrast to intimal mineralization, medial vascular mineralization is mostly found in young and middle-aged individuals with mineral ion dysregulation, as noted in patients with chronic kidney disease (CKD). Medial mineralization affects the elastic lamina of large- and medium-sized arteries, and exerts injurious effects on vascular smooth muscle cells (VSMCs). Medial vascular mineralization can occur in presence or absence of intimal mineralization. In the advanced stages of patients with CKD, a combination of both forms of vascular mineralization has been reported [3].

Medial vascular mineralization is a multievent process, actively regulated by a selected group of molecules [4]. The triggering events of vascular mineralization is not yet clearly defined, although it is likely that an initial cytotoxic insult to the cellar components of the vessels can initiate the process, followed by the abnormal accumulation of hydroxyapatite crystals in and around the injured area, a process believed to be mediated by the mineralization-promoting factors, which include, hyperphosphatemia, hypercalcemia and hypervitaminosis D [5–8]. In fact, an association between serum phosphate levels and vascular mineralization has been reported in the general population; in the CARDIA (Coronary Artery Risk Development in Young Adults) study on 3,015 individuals [9], a 52% higher risk of coronary artery mineralization was documented in participants after 15 years later, whose initial serum phosphate levels were >3.9 mg/dl, as compared to the participants with a serum phosphate levels <3.3 mg/dl [9].

In addition to the mineralization-promoting factors, reduction of the vascular mineralization repressor factors, including ANK (a transmembrane protein), and matrix Gla protein (MGP) can also facilitate the genesis of vascular mineralization [10–12]. For instance, patients with Keutel syndrome with MGP deficiency, and mice without MGP activities develop ectopic mineralization [13, 14]. The calcium/hydroxyapatite-chelating ability of Gla residues disappears in the absence of MGP activity, causing ectopic accumulation of mineralized materials in various organs, as demonstrated in both human and animal studies [13, 15]. It is important to note that reduced activity of mineralization repressor factors can induce ectopic mineralization even at normal calcium and phosphate levels; a few important inducers and inhibitors of vascular mineralization are listed in table 1. Identification of such inducers and inhibitors raise the possibility that, if manipulated in early stages, the progression of vascular mineralization might be delayed or even reversed, particularly in patients with CKD. Of particular significance, hyperphosphatemia in CKD patients undergoing dialysis treatment correlates

Table 1. Partial list of known inducers and inhibitors of vascular calcification

Inducers	Inhibitors
Vitamin D	Pyrophosphate
Uremic toxins	PTH/PTHrP
TNAP	Osteoprotogerin
TGFβ1	Osteopontin
RUNX2	Omega-3 fatty acid
Phosphate	NPP-1
Pit-1	Matrix Gla protein
Osteocalcin	Magnesium
MSX2	Klotho
Leptin	Insulin/IGF-1
LDL	HDL
Inflammatory cytokines	Fetuin-A
Hyperparathyroidism	BMP-7
Calcium	ANK
BMP-2/4	Adiponectin

Modified from an earlier publication [83]. ANK = Transmembrane protein; BMP = bone morphogenetic protein; LDL = low-density lipoprotein; HDL = high-density lipoprotein; IGF-1 = insulin-like growth factor-1; NPP-1 = pyrophosphatase/phosphodiesterase; PTH = parathyroid hormone; PTHrP = PTH-related peptide; RUNX2 = runt-related transcription factor 2; TGF-β_1 = transforming growth factor-β_1; TNAP = tissue nonspecific alkaline phosphatase.

with the occurrence of vascular mineralization, and, more importantly, reducing phosphate burden in these patients can attenuate progression of vascular mineralization [16]. There are numerous excellent reviews available on various molecular and clinical aspects of vascular mineralization [3, 5, 17, 18], and the aim of this brief article is to focus on the role of phosphate in vascular mineralization in CKD.

Chronic Kidney Disease and Vascular Mineralization

Almost 50% of deaths in the advanced stages of patients with CKD are related to cardiovascular anomalies [19–21]. Vascular mineralization, particularly affecting the coronary artery, is strongly associated with the mortality of CKD patients receiving hemodialysis treatment, and such an association is independent of established atherogenic markers [22]. The rate of coronary artery mineralization in CKD patients undergoing hemodialysis treatment is as high as 5 times more than in nondialysis CKD patients [23]. In addition, more than 50% of CKD patients receiving hemodialysis treatment showed features of cardiac valvular calcification [23]. It, however, needs to be mentioned that an increased occurrence of vascular mineralization in patients with CKD is not always associated with hemodialysis treatment, and, as such, vascular pathology has also been reported in predialysis era patients [24, 25].

Using standard radiographic studies, vascular mineralization could be detected in about 60% of patients with advanced stages of CKD [26]. The availability of electron beam and multidetector-row computed tomography imaging system has helped clinicians, not only to detect vascular mineralization in earlier stages, but also provided the tool to monitor the in vivo progression of vascular pathology, and thereby, predicting the consequences of vascular mineralization in patients with CKD. Using such electron beam computed tomography system, an increased occurrence of vascular mineralization affecting coronary arteries has been reported among CKD patients with diabetes, when compared with the nondiabetic CKD patients (94 vs. 59%) [27]. It is, however, needed to mention that the use of electron beam computed tomography is restricted, as most of the dialysis centers do not have the access to such a diagnostic tool. Another important aspect to note is that ectopic vascular mineralization can occur even in general population, but it appears much earlier in patients with CKD [22, 28]. Of relevance, an increased serum level of phosphate, even without kidney disease, has shown to be related to cardiovascular anomalies and death [29].

Vascular mineralization and skeletal mineralization share certain identical molecular events [30]. For instance, analyzing the calcium compound taken from arterial mineralization of CKD patients demonstrated similar hydroxyapatite crystals content, as found in skeleton [31, 32]. It is presumed that altered mineral ion balance can exert cytotoxic effects on both endothelial and VSMCs and facilitate phenotypic transformation of VSMCs into skeletal cells. Once VSMCs are transformed to osteogenic cells, the process of ectopic mineralization leads to the development of vascular mineralization; bone morphogenetic proteins, RUNX2, MSX2 and the activation of WNT signaling are linked to such cellular transdifferentiation processes [7, 33]. In addition, vitamin D can also induce transdifferentiation of VSCMs into the osteoblast-like cells, and thereby promote genesis of vascular mineralization [34–36] (fig. 1). It is important to note that, despite survival benefits of using vitamin D analogs in treating CKD patients [37], such a therapeutic approach often poses a dilemma, as studies have linked vitamin D treatment to subsequent vascular mineralization [25, 30, 38]. However, we have recently found that lowering the burden of phosphate toxicity can reduce vascular mineralization in the experimental mouse models, even in the presence of high serum calcium and vitamin D levels [39].

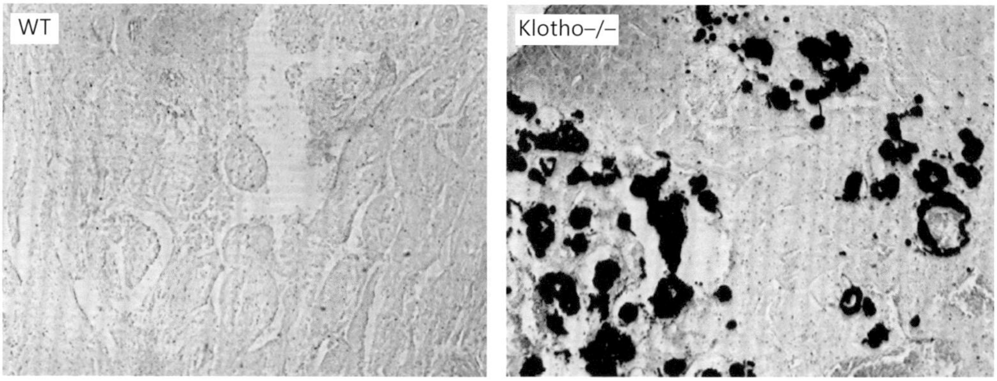

Fig. 2. Von Kossa staining of renal sections prepared from wild-type (WT), and *klotho–/–*, mice. Note that the extensive renovascular calcification (black staining) in the kidney obtained from *klotho–/–* mice with extremely high serum levels of phosphate [82]. No such renal calcification is found in aged-matched WT mice. ×20.

icking CKD suggest that high vitamin D and calcium are not able to induce vascular mineralization when serum phosphate balance is restored within normal range. Further studies will determine whether controlling phosphate balance may increase the safely margins for vitamin D treatment in patients with CKD.

Conclusion

Maintaining a normal phosphate balance in CKD patients to reduce or delay vascular mineralization is the key to their effective management. Usually, the control of phosphate balance in CKD patients is initiated only after the appearance of hyperphosphatemia. However, the fatal consequences of phosphate toxicity in the patients with CKD warrant modification of such a routine therapeutic approach: reduction of phosphate intake by dietary restriction and the use of phosphate binders and other available therapeutics should be implemented ahead of the appearance of hyperphosphatemia [67]. Of relevance, dietary phosphate restriction has been shown to reduce the production of PTH in experimental models of early stage CKD, irrespective of serum calcium and calcitriol concentrations [68–70]. It should be mentioned that dietary phosphate restriction alone is not sufficient to lower phosphate burden in later stages of CKD, and thus the use of phosphate lowering drugs become necessary to remove ingested phosphate from the intestinal lumen to reduce its absorption. Unfortunately, some of the intestinal phosphate-chelating agents are not safe for prolonged use, as calcium-containing phosphate binders can also provoke vascular mineralization and calciphylaxis. Furthermore, despite that aluminum- and magnesium-based phosphate binders are effective in lowering phosphate levels, these compounds are not safe enough for long-term use due to their toxicity. To reduce such toxicity, nonabsorbable polymers are being utilized. For instance, sevelamer hydrochloride has shown to effectively reduce phosphate burden without inducing hypercalcemia in CKD patients [71]. More importantly, some studies claimed that the use of sevelamer compound could reduce the progression of vascular mineralization [72]. It is, however, necessary to point out that careful analysis of sevelamer-treated dialysis patients failed to show any significant reduction of overall mortality [73–75], implicating a need for further exploration to manufacture an effective phosphate-lowering drug. One of the major improvements in designing the therapeutic strategy for patients with CKD is realization of the importance of phosphate toxicity in the overall survival of these patients [76, 77]. Such an awareness [78] and realization has helped in focusing research towards generating effective and less toxic phosphate-lowering agents that have enormous therapeutic promise.

Acknowledgements

Part of the research work is supported by the grant (R01-DK077276 to M.S.R.) provided by the NIH (NIDDK). Thanks are due to Razzaque lab members, including, Mutsuko Ohnishi (MD, PhD), Junko Akiyoshi (MD), Satoko Osuka (MD), Yongguen Hong (PhD), Khadijah Turkistani (BDS) and Ismai Eddafali of Harvard School of Dental Medicine (Boston) for their original research works, some of which formed the basis of this book chapter. I would like to express my gratitude to Syed Rafi (PhD) for critically reading this chapter and offering useful suggestions.

References

1 Stary HC: Lipid and macrophage accumulations in arteries of children and the development of atherosclerosis. Am J Clin Nutr 2000; 72:1297S–1306S.

2 Shroff RC, Shanahan CM: The vascular biology of calcification. Semin Dial 2007;20:103–109.

3 Giachelli CM: Vascular calcification mechanisms. J Am Soc Nephrol 2004;15:2959–2964.

4 Guerin AP, London GM, Marchais SJ, Metivier F: Arterial stiffening and vascular calcifications in end-stage renal disease. Nephrol Dial Transplant 2000;15:1014–1021.

5 Moe SM, Chen NX: Mechanisms of vascular calcification in chronic kidney disease. J Am Soc Nephrol 2008;19:213–216.

6 El-Abbadi MM, Pai AS, Leaf EM, Yang HY, Bartley BA, Quan KK, Ingalls CM, Liao HW, Giachelli CM: Phosphate feeding induces arterial medial calcification in uremic mice: role of serum phosphorus, fibroblast growth factor-23, and osteopontin. Kidney Int 2009; 75:1297–1307.

7 Towler DA, Shao JS, Cheng SL, Pingsterhaus JM, Loewy AP: Osteogenic regulation of vascular calcification. Ann NY Acad Sci 2006; 1068:327–333.

8 Razzaque MS, St-Arnaud R, Taguchi T, Lanske B: FGF-23, vitamin D and calcification: the unholy triad. Nephrol Dial Transplant 2005;20:2032–2035.
9 Foley RN, Collins AJ, Herzog CA, Ishani A, Kalra PA: Serum phosphorus levels associate with coronary atherosclerosis in young adults. J Am Soc Nephrol 2009;20:397–404.
10 Mori K, Shioi A, Jono S, Nishizawa Y, Morii H: Expression of matrix Gla protein (MGP) in an in vitro model of vascular calcification. FEBS Lett 1998;433:19–22.
11 Jono S, Ikari Y, Vermeer C, Dissel P, Hasegawa K, Shioi A, Taniwaki H, Kizu A, Nishizawa Y, Saito S: Matrix Gla protein is associated with coronary artery calcification as assessed by electron-beam computed tomography. Thromb Haemost 2004;91:790–794.
12 Gheduzzi D, Boraldi F, Annovi G, DeVincenzi CP, Schurgers LJ, Vermeer C, Quaglino D, Ronchetti IP: Matrix Gla protein is involved in elastic fiber calcification in the dermis of pseudoxanthoma elasticum patients. Lab Invest 2007;87:998–1008.
13 Hur DJ, Raymond GV, Kahler SG, Riegert-Johnson DL, Cohen BA, Boyadjiev SA: A novel MGP mutation in a consanguineous family: review of the clinical and molecular characteristics of Keutel syndrome. Am J Med Genet A 2005;135:36–40.
14 El-Maadawy S, Kaartinen MT, Schinke T, Murshed M, Karsenty G, McKee MD: Cartilage formation and calcification in arteries of mice lacking matrix Gla protein. Connect Tissue Res 2003;44(suppl 1):272–278.
15 Luo G, Ducy P, McKee MD, Pinero GJ, Loyer E, Behringer RR, Karsenty G: Spontaneous calcification of arteries and cartilage in mice lacking matrix Gla protein. Nature 1997;386: 78–81.
16 Raggi P, Bellasi A: Clinical assessment of vascular calcification. Adv Chron Kid Dis 2007; 14:37–43.
17 Ossareh S: Vascular calcification in chronic kidney disease: mechanisms and clinical implications. Iran J Kidney Dis 2011;5:285–299.
18 Mizobuchi M, Towler D, Slatopolsky E: Vascular calcification: the killer of patients with chronic kidney disease. J Am Soc Nephrol 2009;20:1453–1464.
19 Blacher J, Guerin AP, Pannier B, Marchais SJ, London GM: Arterial calcifications, arterial stiffness, and cardiovascular risk in end-stage renal disease. Hypertension 2001;38:938–942.
20 Culleton BF, Hemmelgarn BR: Is chronic kidney disease a cardiovascular disease risk factor? Semin Dial 2003;16:95–100.
21 Levin A, Stevens L, McCullough PA: Cardiovascular disease and the kidney. Tracking a killer in chronic kidney disease. Postgrad Med 2002;111:53–60.
22 London GM, Guerin AP, Marchais SJ, Metivier F, Pannier B, Adda H: Arterial media calcification in end-stage renal disease: impact on all-cause and cardiovascular mortality. Nephrol Dial Transplant 2003;18:1731–1740.
23 Braun J, Oldendorf M, Moshage W, Heidler R, Zeitler E, Luft FC: Electron beam computed tomography in the evaluation of cardiac calcification in chronic dialysis patients. Am J Kidney Dis 1996;27:394–401.
24 Pollak VE, Schneider AF, Freund G, Kark RM: Chronic renal disease with secondary hyperparathyroidism. AMA Arch Intern Med 1959;103:200–218.
25 Parfitt AM: Soft-tissue calcification in uremia. Arch Intern Med 1969;124:544–556.
26 Meema HE, Oreopoulos DG: Morphology, progression, and regression of arterial and periarterial calcifications in patients with end-stage renal disease. Radiology 1986;158: 671–677.
27 Merjanian R, Budoff M, Adler S, Berman N, Mehrotra R: Coronary artery, aortic wall, and valvular calcification in nondialyzed individuals with type 2 diabetes and renal disease. Kidney Int 2003;64:263–271.
28 Goodman WG, Goldin J, Kuizon BD, Yoon C, Gales B, Sider D, Wang Y, Chung J, Emerick A, Greaser L, Elashoff RM, Salusky IB: Coronary-artery calcification in young adults with end-stage renal disease who are undergoing dialysis. N Engl J Med 2000;342:1478–1483.
29 Chonchol M, Dale R, Schrier RW, Estacio R: Serum phosphorus and cardiovascular mortality in type 2 diabetes. Am J Med 2009;122: 380–386.
30 Mizobuchi M, Ogata H, Koiwa F, Kinugasa E, Akizawa T: Vitamin d and vascular calcification in chronic kidney disease. Bone 2009; 45(suppl 1):S26–S29.
31 Contiguglia SR, Alfrey AC, Miller NL, Runnells DE, Le Geros RZ: Nature of soft tissue calcification in uremia. Kidney Int 1973;4: 229–235.

32 Reid JD, Andersen ME: Medial calcification (whitlockite) in the aorta. Atherosclerosis 1993;101:213–224.
33 Al-Aly Z, Shao JS, Lai CF, Huang E, Cai J, Behrmann A, Cheng SL, Towler DA: Aortic Msx2-Wnt calcification cascade is regulated by TNF-alpha-dependent signals in diabetic Ldlr–/– mice. Arterioscler Thromb Vasc Biol 2007;27:2589–2596.
34 Tukaj C, Kubasik-Juraniec J, Kraszpulski M: Morphological changes of aortal smooth muscle cells exposed to calcitriol in culture. Med Sci Monit 2000;6:668–674.
35 Rebsamen MC, Sun J, Norman AW, Liao JK: 1alpha,25-Dihydroxyvitamin D_3 induces vascular smooth muscle cell migration via activation of phosphatidylinositol 3-kinase. Circ Res 2002;91:17–24.
36 Tukaj C: Enhanced proliferation of aortal smooth muscle cells treated by 1,25(OH)2D3 in vitro coincides with impaired formation of elastic fibres. Int J Exp Pathol 2008;89:117–124.
37 Pilz S, Iodice S, Zittermann A, Grant WB, Gandini S: Vitamin D status and mortality risk in CKD: a meta-analysis of prospective studies. Am J Kidney Dis 2011;58:374–382.
38 Kalantar-Zadeh K, Kovesdy CP: Clinical outcomes with active versus nutritional vitamin D compounds in chronic kidney disease. Clin J Am Soc Nephrol 2009;4:1529–1539.
39 Ohnishi M, Nakatani T, Lanske B, Razzaque MS: In vivo genetic evidence for suppressing vascular and soft-tissue calcification through the reduction of serum phosphate levels, even in the presence of high serum calcium and 1,25-dihydroxyvitamin D levels. Circ Cardiovasc Genet 2009;2:583–590.
40 Gaasbeek A, Meinders AE: Hypophosphatemia: an update on its etiology and treatment. Am J Med 2005;118:1094–1101.
41 ADHR_Consortium: Autosomal dominant hypophosphataemic rickets is associated with mutations in FGF23. The ADHR consortium. Nat Genet 2000;26:345–348.
42 Shimada T, Mizutani S, Muto T, Yoneya T, Hino R, Takeda S, Takeuchi Y, Fujita T, Fukumoto S, Yamashita T: Cloning and characterization of FGF23 as a causative factor of tumor-induced osteomalacia. Proc Natl Acad Sci USA 2001;98:6500–6505.
43 Bai X, Miao D, Li J, Goltzman D, Karaplis AC: Transgenic mice overexpressing human fibroblast growth factor 23 (R176Q) delineate a putative role for parathyroid hormone in renal phosphate wasting disorders. Endocrinology 2004;145:5269–5279.
44 Kurosu H, Ogawa Y, Miyoshi M, Yamamoto M, Nandi A, Rosenblatt KP, Baum MG, Schiavi S, Hu MC, Moe OW, Kuro-o M: Regulation of fibroblast growth factor-23 signaling by klotho. J Biol Chem 2006;281:6120–6123.
45 Goetz R, Beenken A, Ibrahimi OA, Kalinina J, Olsen SK, Eliseenkova AV, Xu C, Neubert TA, Zhang F, Linhardt RJ, Yu X, White KE, Inagaki T, Kliewer SA, Yamamoto M, Kurosu H, Ogawa Y, Kuro-o M, Lanske B, Razzaque MS, Mohammadi M: Molecular insights into the klotho-dependent, endocrine mode of action of fibroblast growth factor 19 subfamily members. Mol Cell Biol 2007;27:3417–3428.
46 Goetz R, Nakada Y, Hu MC, Kurosu H, Wang L, Nakatani T, Shi M, Eliseenkova AV, Razzaque MS, Moe OW, Kuro-o M, Mohammadi M: Isolated C-terminal tail of FGF23 alleviates hypophosphatemia by inhibiting FGF23-FGFR-klotho complex formation. Proc Natl Acad Sci USA 2010;107:407–412.
47 Urakawa I, Yamazaki Y, Shimada T, Iijima K, Hasegawa H, Okawa K, Fujita T, Fukumoto S, Yamashita T: Klotho converts canonical fgf receptor into a specific receptor for FGF23. Nature 2006;444:770–774.
48 Medici D, Razzaque MS, Deluca S, Rector TL, Hou B, Kang K, Goetz R, Mohammadi M, Kuro OM, Olsen BR, Lanske B: FGF-23-klotho signaling stimulates proliferation and prevents vitamin D-induced apoptosis. J Cell Biol 2008;182:459–465.
49 Nabeshima Y: The discovery of alpha-klotho and FGF23 unveiled new insight into calcium and phosphate homeostasis. Cell Mol Life Sci 2008;65:3218–3230.
50 Razzaque MS: FGF23-mediated regulation of systemic phosphate homeostasis: Is klotho an essential player? Am J Physiol Renal Physiol 2009;296:F470–F476.
51 Nakatani T, Sarraj B, Ohnishi M, Densmore MJ, Taguchi T, Goetz R, Mohammadi M, Lanske B, Razzaque MS: In vivo genetic evidence for klotho-dependent, fibroblast growth factor 23 (FGF23)-mediated regulation of systemic phosphate homeostasis. FASEB J 2009;23:433–441.

52 Kuro-o M, Matsumura Y, Aizawa H, Kawaguchi H, Suga T, Utsugi T, Ohyama Y, Kurabayashi M, Kaname T, Kume E, Iwasaki H, Iida A, Shiraki-Iida T, Nishikawa S, Nagai R, Nabeshima YI: Mutation of the mouse klotho gene leads to a syndrome resembling ageing. Nature 1997;390:45–51.
53 Razzaque MS: Phosphate toxicity: new insights into an old problem. Clin Sci (Lond) 2011;120:91–97.
54 Razzaque MS: Osteo-renal regulation of systemic phosphate metabolism. IUBMB Life 2011;63:240–247.
55 Razzaque MS: The FGF23 klotho axis: endocrine regulation of phosphate homeostasis. Nat Rev Endocrinol 2009;5:611–619.
56 Razzaque MS, Sitara D, Taguchi T, St-Arnaud R, Lanske B: Premature aging-like phenotype in fibroblast growth factor 23 null mice is a vitamin D-mediated process. FASEB J 2006;20:720–722.
57 Sitara D, Kim S, Razzaque MS, Bergwitz C, Taguchi T, Schuler C, Erben RG, Lanske B: Genetic evidence of serum phosphate-independent functions of FGF-23 on bone. PLoS Genet 2008;4:e1000154.
58 Ohnishi M, Razzaque MS: Dietary and genetic evidence for phosphate toxicity accelerating mammalian aging. FASEB J 2010;24: 3562–3571.
59 Nabeshima Y: Toward a better understanding of klotho. Sci Aging Knowledge Environ 2006;2006:pe11.
60 Ohnishi M, Kato S, Razzaque MS: Genetic induction of phosphate toxicity significantly reduces the survival of hypercholesterolemic obese mice. Biochem Biophys Res Commun 2011;415:434–438.
61 Ohnishi M, Razzaque MS: Osteo-renal cross-talk and phosphate metabolism by fgf23-klotho system. Contrib Nephrol. Basel, Karger, 2013, vol 180, pp 1–13.
62 Ohnishi M, Kato S, Akiyoshi J, Atfi A, Razzaque MS: Dietary and genetic evidence for enhancing glucose metabolism and reducing obesity by inhibiting klotho functions. FASEB J 2011;25:2031–2039.
63 Voormolen N, Noordzij M, Grootendorst DC, Beetz I, Sijpkens YW, van Manen JG, Boeschoten EW, Huisman RM, Krediet RT, Dekker FW: High plasma phosphate as a risk factor for decline in renal function and mortality in pre-dialysis patients. Nephrol Dial Transplant 2007;22:2909–2916.
64 Wang TJ, Pencina MJ, Booth SL, Jacques PF, Ingelsson E, Lanier K, Benjamin EJ, D'Agostino RB, Wolf M, Vasan RS: Vitamin D deficiency and risk of cardiovascular disease. Circulation 2008;117:503–511.
65 Shoji T, Shinohara K, Kimoto E, Emoto M, Tahara H, Koyama H, Inaba M, Fukumoto S, Ishimura E, Miki T, Tabata T, Nishizawa Y: Lower risk for cardiovascular mortality in oral 1alpha-hydroxy vitamin D_3 users in a haemodialysis population. Nephrol Dial Transplant 2004;19:179–184.
66 Wolisi GO, Moe SM: The role of vitamin D in vascular calcification in chronic kidney disease. Semin Dial 2005;18:307–314.
67 Osuka S, Razzaque MS: Can features of phosphate toxicity appear in normophosphatemia? J Bone Miner Metab 2012;30:10–18.
68 Graciolli FG, Neves KR, dos Reis LM, Graciolli RG, Noronha IL, Moyses RM, Jorgetti V: Phosphorus overload and PTH induce aortic expression of runx2 in experimental uraemia. Nephrol Dial Transplant 2009;24:1416–1421.
69 Guntupalli J, Matthews B, Carlin B, Bourke E: Effect of acute hypercapnia on pth-stimulated phosphaturia in dietary Pi-deprived rat. Am J Physiol 1987;253:F34–F40.
70 Murer H, Evers C, Stoll R, Kinne R: The effect of parathyroid hormone (PTH) and dietary phosphate on the sodium-dependent phosphate transport system located in the rat renal brush border membrane. Curr Probl Clin Biochem 1977;8:455–462.
71 Chertow GM, Burke SK, Dillon MA, Slatopolsky E: Long-term effects of sevelamer hydrochloride on the calcium x phosphate product and lipid profile of haemodialysis patients. Nephrol Dial Transplant 1999;14: 2907–2914.
72 Raggi P, Bommer J, Chertow GM: Valvular calcification in hemodialysis patients randomized to calcium-based phosphorus binders or sevelamer. J Heart Valve Dis 2004;13: 134–141.
73 Tonelli M, Wiebe N, Culleton B, Lee H, Klarenbach S, Shrive F, Manns B: Systematic review of the clinical efficacy and safety of sevelamer in dialysis patients. Nephrol Dial Transplant 2007;22:2856–2866.

74 Barreto DV, Barreto Fde C, de Carvalho AB, Cuppari L, Draibe SA, Dalboni MA, Moyses RM, Neves KR, Jorgetti V, Miname M, Santos RD, Canziani ME: Phosphate binder impact on bone remodeling and coronary calcification – results from the BRIC study. Nephron Clin Pract 2008;110:c273–c283.
75 Qunibi W, Moustafa M, Muenz LR, He DY, Kessler PD, Diaz-Buxo JA, Budoff M: A 1-year randomized trial of calcium acetate versus sevelamer on progression of coronary artery calcification in hemodialysis patients with comparable lipid control: the Calcium Acetate Renagel Evaluation-2 (CARE-2) study. Am J Kidney Dis 2008;51:952–965.
76 Block GA, Hulbert-Shearon TE, Levin NW, Port FK: Association of serum phosphorus and calcium X phosphate product with mortality risk in chronic hemodialysis patients: A national study. Am J Kidney Dis 1998;31: 607–617.
77 Block GA, Raggi P, Bellasi A, Kooienga L, Spiegel DM: Mortality effect of coronary calcification and phosphate binder choice in incident hemodialysis patients. Kidney Int 2007;71:438–441.
78 Shutto Y, Shimada M, Kitajima M, Yamabe H, Razzaque MS: Lack of awareness among future medical professionals about the risk of consuming hidden phosphate-containing processed food and drinks. PLoS One 2011; 6:e29105.
79 Wada T, McKee MD, Steitz S, Giachelli CM: Calcification of vascular smooth muscle cell cultures: inhibition by osteopontin. Circ Res 1999;84:166–178.
80 Son BK, Kozaki K, Iijima K, Eto M, Nakano T, Akishita M, Ouchi Y: Gas6/Axl-PI3k/Akt pathway plays a central role in the effect of statins on inorganic phosphate-induced calcification of vascular smooth muscle cells. Eur J Pharmacol 2007;556:1–8.
81 Son BK, Kozaki K, Iijima K, Eto M, Kojima T, Ota H, Senda Y, Maemura K, Nakano T, Akishita M, Ouchi Y: Statins protect human aortic smooth muscle cells from inorganic phosphate-induced calcification by restoring Gas6-Axl survival pathway. Circ Res 2006;98: 1024–1031.
82 Nakatani T, Ohnishi M, Razzaque MS: Inactivation of klotho function induces hyperphosphatemia even in presence of high serum fibroblast growth factor 23 levels in a genetically engineered hypophosphatemic (HYP) mouse model. FASEB J 2009;23:3702–3711.
83 Razzaque MS: The dualistic role of vitamin D in vascular calcifications. Kidney Int 2011;79: 708–714.

Mohammed S. Razzaque, MBBS, PhD
Department of Oral Medicine, Infection and Immunity
Harvard School of Dental Medicine, Research and Education Building
Room 304, 190 Longwood Avenue, Boston, MA 02115 (USA)
E-Mail mrazzaque@hms.harvard.edu, debpur@gmail.com

Razzaque MS (ed): Phosphate and Vitamin D in Chronic Kidney Disease.
Contrib Nephrol. Basel, Karger, 2013, vol 180, pp 86–97 (DOI: 10.1159/000346786)

Vitamin D and Type II Sodium-Dependent Phosphate Cotransporters

Shinsuke Kido · Ichiro Kaneko · Sawako Tatsumi · Hiroko Segawa · Ken-ichi Miyamoto

Department of Molecular Nutrition, University of Tokushima Graduate School, Tokushima, Japan

Abstract

The type II sodium-dependent Pi (NaPi) cotransporters (NaPi-IIa, NaPi-IIb and NaPi-IIc) contribute to renal and intestinal Pi absorption. 1,25-Dihydroxyvitamin D [$1,25(OH)_2D_3$] is an important factor for NaPi-II transporters in the small intestine and kidney. In a previous study, low levels of $1,25(OH)_2D_3$ appeared to suppress the expression of renal NaPi cotransporters. We identified a functional vitamin D receptor-responsive element in the human NaPi-IIa and NaPi-IIc genes in renal epithelial cells. In an analysis of vitamin D receptor (VDR)-null mice, we observed early onset of hypophosphatemia. The cause of the hypophosphatemia in VDR-null mice before weaning appeared to be increased plasma parathyroid hormone (PTH) levels during the suckling periods. A rescue diet (high calcium diet) decreased plasma PTH levels in VDR-null mice. The reduced plasma PTH levels normalized the renal Npt2a and Npt2c protein levels in weanling animals. Thus, the dietary intervention completely normalized the expression of the renal Pi transporters (Npt2a/Npt2c) in VDR-null mice, suggesting that the lack of VDR activity was not the cause of the impaired renal Pi reabsorption. In suckling animals, $1,25(OH)_2D_3$ may be essential for the prevention of the phosphaturic action of PTH. In adult animals, $1,25(OH)_2D_3$ is thought to be an important factor for posttranscriptional regulation of the Npt2b gene in the small intestine. Fibroblast growth factor 23 (FGF23) is a novel phosphaturic factor that influences vitamin D metabolism and renal reabsorption of Pi. We characterized the role of the VDR in the action of FGF23 using VDR-null mice. FGF23 reduced renal Pi transport and 25-hydroxyvitamin D 1a-hydroxylase levels by a mechanism that was independent of the VDR. By contrast, the induction of 25-hydroxyvitamin D 24-hydroxylase and the reduction in serum $1,25(OH)_2D_3$ levels induced by FGF23 were dependent on the VDR. Thus, the VDR is not essential for the phosphaturic action of FGF23, but is essential for control of the plasma $1,25(OH)_2D_3$ level. Moreover, FGF23 reduces intestinal NaPi transport activity and Npt2b protein levels by a mechanism that is dependent on the VDR. Klotho functions as a co-receptor for FGF23 and is increased by $1,25(OH)_2D_3$. Klotho induces phosphaturia by inhibiting the renal NaPi-IIa transporter. In this review, we discuss the roles of $1,25(OH)_2D_3$/VDR in the regulation of renal type II NaPi cotransporters in the kidney and small intestine.

The balance between phosphate (Pi) absorption from the intestine, mobilization from bone, and excretion via the kidneys in the form of urine determines the final levels of plasma Pi [1–3]. In the kidneys, up to 65–70% of the filtered Pi is reabsorbed via the proximal tubules, in which the Na^+-dependent Pi (NaPi) transport system within the brush-border membrane (BBM) mediates the rate-limiting step in the overall Pi reabsorption process [1–3]. Pi reabsorption within the kidneys is primarily mediated by type II sodium-dependent Pi cotransporters (NaPi-II/Npt2) that belong to the SLC34 gene family. SLC34A1 (NaPi-IIa/Npt2a), SLC34A2 (NaPi-IIb/Npt2b), and SLC34A3 (NaPi-IIc/Npt2c) are the most well-characterized family members with regard to their function and regulation [1–3]. Regarding the functional properties of sodium-dependent Pi cotransport, the SLC34 family comprises the electrogenic NaPi-IIa/b and the electroneutral NaPi-IIc, which display Na^+:Pi cotransport stoichiometries of 3:1 and 2:1, respectively [2]. NaPi-IIa and NaPi-IIc are localized at the apical membrane of proximal tubular cells, while NaPi-IIb is mainly expressed in the lungs and small intestine [1–3]. In rats and mice, NaPi-IIa plays the most important role in renal Pi reabsorption (being responsible for 60–70%), while NaPi-IIc is thought to have the most important regulatory role in weaning animals [4–6]. In addition, NaPi-IIb is thought to be a major Na-dependent Pi cotransporter in the small intestine [7].

Most actions of vitamin D require activation of the vitamin D receptor (VDR) by 1,25-dihydroxyvitamin D [$1,25(OH)_2D_3$], the most active vitamin D metabolite and a potent steroid hormone [8]. The type II NaPi transporters are classically regulated via the vitamin D endocrine system and parathyroid hormone (PTH). In addition, phosphatonins, such as fibroblast growth factor 23 (FGF23) and klotho, have emerged as regulators of Pi homeostasis [9]. FGF23 is a novel phosphaturic factor that influences vitamin D metabolism and renal reabsorption of Pi [9, 10]. Klotho is recognized as a cofactor that is mandatory for FGF23 action [11, 12]. FGF23 mediates its action in the kidney through an FGF receptor/Klotho complex that downregulates NaPi-IIa and NaPi-IIc expression in the proximal tubules. FGF23 also inhibits $1,25(OH)_2D_3$ production [9, 10, 13]. By contrast, PTH stimulates Pi excretion and $1,25(OH)_2D_3$ synthesis in the kidneys, as well as FGF23 production in bones [9, 10]. The vitamin D endocrine system regulates body Pi homeostasis through the actions of PTH and the FGF23/klotho system [9, 10].

The adverse events of hyperphosphatemia in chronic kidney disease, secondary hyperparathyroidism, calcium and vitamin D derangements, vascular calcification and metabolic bone disorders arise from adaptations of nephrons attempting to preserve Pi homeostasis [14]. Vitamin D insufficiency is commonly observed in patients with chronic kidney disease. The regulation of renal Pi reabsorption by the vitamin D endocrine system is important for understanding Pi metabolism in chronic kidney disease [14]. In this review, we discuss the roles of $1,25(OH)_2D_3$ in the regulation of renal and intestinal type II NaPi transporters.

Regulation of Pi Reabsorption by $1,25(OH)_2D_3$

Vitamin D plays a central role in modulating Pi homoeostasis and uptake in the small intestine and kidneys. It is possible that inappropriately low levels of $1,25(OH)_2D_3$ may suppress the expression of renal and intestinal NaPi cotransporters [1–3]. Although the profound effects of $1,25(OH)_2D_3$ on the metabolism and homeostasis of Pi are well established, conflicting findings surrounding the direct effects of $1,25(OH)_2D_3$ on Pi reabsorption in proximal tubules have been reported [15–17]. This may have arisen because the vitamin D status is closely associated with alterations in plasma calcium, PTH and FGF23 levels. In addition, the effect of $1,25(OH)_2D_3$ on Pi reabsorption has been suggested to be related to changes in lipid characteristics [18, 19].

In a previous study, we demonstrated the action of $1,25(OH)_2D_3$ on renal NaPi-IIa gene expression [15]. In vitamin D-deficient rats, we found that the amounts of NaPi-IIa protein and mRNA were decreased, compared with control rats, in the juxtamedullary kidney cortex, but not the superficial cortex. Administration of $1,25(OH)_2D_3$ to the vitamin D-deficient rats increased the initial rate of Pi uptake as well as the amounts of NaPi-IIa mRNA and protein in the juxtamedullary cortex [15]. The transcriptional activity of a luciferase reporter plasmid containing the promoter region of the human NaPi-IIa gene was markedly increased by $1,25(OH)_2D_3$ in COS-7 cells expressing human VDR. A deletion and mutation analysis of the NaPi-IIa gene promoter identified the vitamin D-responsive element (VDRE) as the sequence 5′-GGGGCAGCAAGGGCA-3′, comprising nucleotides –1977 to –1963 relative to the transcription start site. This element bound a heterodimer of VDR and retinoid X receptor. Therefore, one mechanism by which vitamin D regulates Pi homeostasis is through modulation of the expression of type IIa NaPi transporter genes in the juxtamedullary kidney cortex [15].

Hereditary hypophosphatemic rickets with hypercalciuria, a mendelian disorder of renal Pi reabsorption, results from the lack of a functional NaPi-IIc protein, and is characterized by severe renal wasting with hypophosphatemia [3]. Recent studies suggested that the NaPi-IIc cotransporter plays an important role in renal Pi reabsorption and may be a key determinant of the plasma Pi concentrations in humans [3]. In addition, the human NPT2c gene promoter has a consensus VDRE sequence and its promoter activity is upregulated by $1,25(OH)_2D_3$ [20]. In a previous study, Masuda et al. [20] observed the effects of a vitamin A-deficient (VAD) diet on Pi homoeostasis and the expression of the Npt2a and Npt2c genes. Renal BBM Pi uptake activity and renal Npt2a and Npt2c expressions were significantly decreased in VAD rats. The transcriptional activity of a luciferase reporter plasmid containing the promoter region of the human Npt2a and Npt2c genes was markedly increased by ATRA and a retinoic acid receptor (RAR)-specific analogue TTNPB {4-[E-2-(5, 6, 7, 8-tetrahydro-5, 5, 8, 8-tetramethyl-2-naphthalenyl)-1-propenyl] benzoic acid} in renal proximal tubular cells overexpressing RARs and retinoid X receptors. Masuda et al. [20] also identified retinoic acid-response elements (RAREs) in both gene promoters. Interestingly, the half-site sequences (5′-GGTTCA-3′: –563 to –558) of 2c-RARE1 overlapped the VDRE in the human Npt2c gene and were functionally important motifs for transcriptional regulation of human Npt2c by ATRA and $1,25(OH)_2D_3$ in both independent or additive actions. These data suggest that $1,25(OH)_2D_3$ has a profound effect on the expression of the human NPT2c gene and that the NaPi-IIa and NaPi-IIc gene promoters have functional VDREs in renal epithelial cells.

Mechanisms of Hypophosphatemia in VDR-Null Mice

$1,25(OH)_2D_3$ binds the VDR and induces changes in target gene expression. To further investigate the role of the vitamin D endocrine system in Pi reabsorption in proximal tubules, we analyzed the levels of NaPi-II (Npt2a and Npt2b) transporters in adult VDR-null mice [21]. VDR-null mice on a normal diet developed rickets, hypocalcemia, hypophosphatemia, elevated plasma $1,25(OH)_2D_3$ and hyperparathyroidism. Adult VDR-null mice exhibited decreased Pi absorption in the intestine with reduced Npt2b expression. The renal Pi reabsorption in VDR-null mice was not decreased, although the plasma PTH levels were markedly increased. We concluded that the hypophosphatemia in VDR-null mice is mainly caused by defective intestinal Pi absorption (jejunum) and is less critical than defective renal Pi reabsorption. Although the Npt2a protein levels were significantly decreased in VDR-null mice, the renal Na-dependent Pi cotransport activity in the BBM vesicles (BBMVs) was similar to that in wild-type mice, indicating that at least in adult mice, Pi reabsorption was not defective in the mutant mice compared with wild-type mice [21] (fig. 1).

By contrast, we found that $1,25(OH)_2D_3$ appears to be important for renal Pi reabsorption in suckling and weaning animals [22]. The fractional reabsorption of Pi was shown to be higher in growing animals than in fully grown animals and humans. VDR-null mice had hyperphosphatemia at 28 days of age. After weaning, VDR-null mice exhibited retarded growth and developed severe rickets,

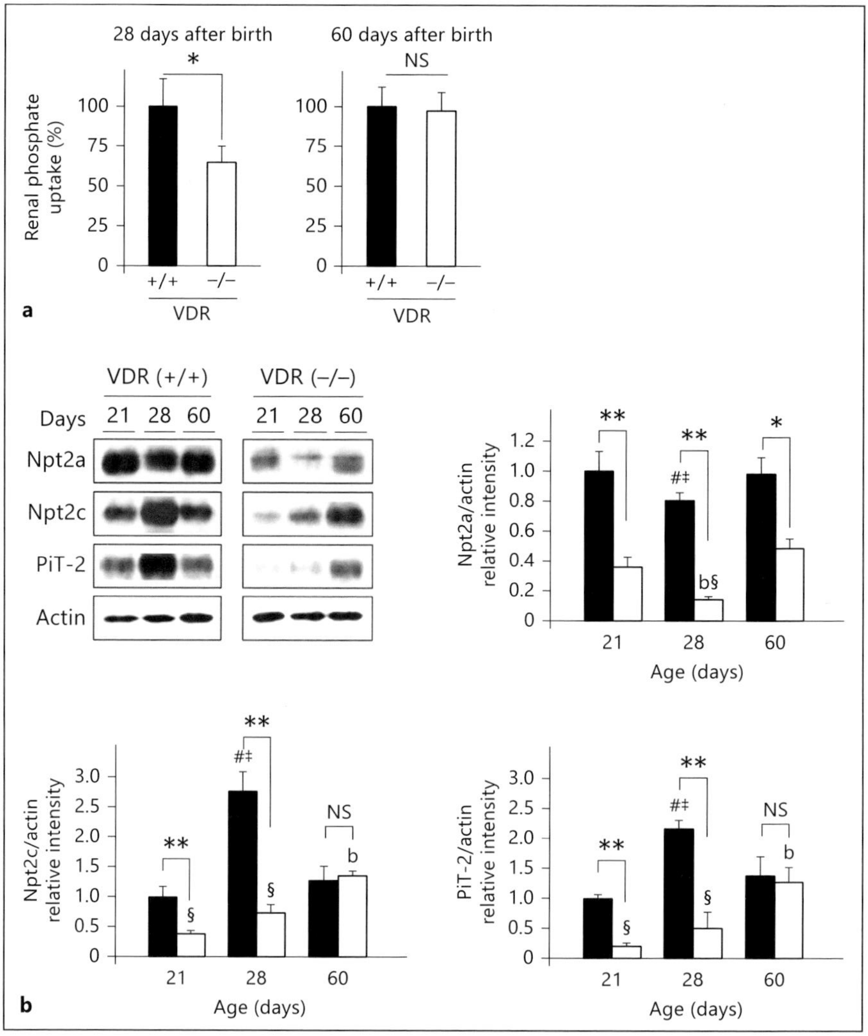

Fig. 1. Renal Na/Pi cotransport activity, and renal Na/Pi cotransporter protein expression in VDR(+/+) and VDR(–/–) mice at 28 and 60 days of age [22]. **a** Na^+-dependent Pi cotransport activity was assessed by measuring Pi uptake (30 s) by renal BBMVs in VDR(+/+) and VDR(–/–) mice at 28 and 60 days of age. Black bars represent VDR(+/+), white bars represent VDR(–/–) mice. Values are shown as mean ± SD. * $p < 0.05$; n = 6–10 mice for each day of age. NS = Not significant. Equilibrium values (180 s) were similar (data not shown). **b** Western blot analysis of the renal Na/Pi cotransporters. BBMVs isolated from kidneys of mice fed the normal diet were loaded into each lane. Data were normalized to actin expression. Relative intensity of Npt2a, Npt2c and PiT-2 expression in renal BBMVs from VDR(+/+) and VDR(–/–) mice. Expression levels in VDR(+/+) mice at 21 days of age were defined as 1.0. Values are shown as mean ± SD. * $p < 0.05$, ** $p < 0.01$ between VDR(+/+) and VDR(–/–) mice at the same age. # $p < 0.05$ compared with VDR(+/+) mice at 21 days of age. ‡ $p < 0.05$ compared with VDR(+/+) mice at 60 days of age. b $p < 0.05$ compared with VDR(–/–) mice at 21 days of age. § $p < 0.05$ compared with VDR(–/–) mice at 60 days of age; n = 4–6 mice for each age group. NS = Not significant.

suggesting that an abnormal plasma Ca/Pi ratio, in addition to secondary hyperparathyroidism, during development triggers the development of rickets. We also observed early onset of hypophosphatemia in VDR-null mice [22]. The cause of the hypophosphatemia in VDR-null mice before weaning appeared to be increased plasma PTH levels during suckling periods [22]. A rescue diet (high calcium diet) decreased the plasma PTH levels in the VDR-null mice. Reduced plasma PTH levels may have normalized the renal Npt2a and Npt2c protein levels in weanling animals (fig. 2). Thus, the dietary intervention completely normalized the expression of the renal Pi transporter (Npt2a/Npt2c/PiT-2) in VDR-null mice, suggesting that the lack of VDR activity was not the cause of the impaired renal Pi reabsorption [22]. Interestingly, PTH did not show a phosphaturic action in adult VDR-null mice. The mechanisms for the PTH resistance of renal Pi reabsorption in adult VDR-null mice are unknown.

Vitamin D and FGF23

Contradictory results showing an increase or decrease in Pi excretion in response to $1,25(OH)_2D_3$ have been reported [15–17]. An increase in plasma $1,25(OH)_2D_3$ concentration (which occurred despite a reduction in PTH) increased renal Pi reabsorption in the proximal tubules, because the reduction of PTH stimulated renal Pi reabsorption. However, when excessive secretion of PTH is observed, $1,25(OH)_2D_3$ promotes a phosphaturic action of PTH. Rubinger et al. [17] reported that vitamin D administration to hyperparathyroid rats with chronic renal failure increased phosphaturia. In addition, Friedlaender et al. [16] showed that administration of $1,25(OH)_2D_3$ reduced the downregulation of renal Pi transport activity and NaPi-IIa in PTH-infused rats. Chronic administration of $1,25(OH)_2D_3$ led to increased PTH-induced phosphaturia despite a decrease in urinary cAMP [17]. In a previous study, we suggested that vitamin D action on renal Pi handling could be indirect, involving altered serum levels of FGF23 [23]. For example, in intact rats, repeated injections of $1,25(OH)_2D_3$ resulted in increased serum levels of FGF23 [23]. In addition, FGF23 downregulated the expression of Npt2a and Npt2c [24], suggesting that the phosphaturic action of $1,25(OH)_2D_3$ in PTH-infused animals may be dependent on the action of FGF23.

The following question arises from these findings. How does the downregulation of vitamin D synthesis by FGF23 have an influence on its phosphaturic action? To address this question, we characterized the role of the VDR in FGF23 action using VDR-null [VDR(–/–)] mice [25]. Injection of FGF23M (naked DNA encoding an R179Q mutant of human FGF23) into VDR(–/–) and wild-type VDR(+/+) mice resulted in elevated serum FGF23 levels, but had no effect

stimulation of intestinal calcium and Pi absorption [32]. Klotho deficiency leads to profound disturbances of mineral metabolism, i.e. hypercalcemia, hyperphosphatemia and vascular calcification [32]. Klotho deficiency leads to excessive formation of $1,25(OH)_2D_3$, contributing to the increase of plasma Pi concentration in Klotho-deficient mice [32]. Recent studies showed that transgenic klotho overexpression or intravenous klotho infusion indeed caused hypophosphatemia and phosphaturia and decreased protein abundance and activity of the major renal transporter NaPi-IIa [33, 34]. FGF23 is not required for klotho-induced phosphaturia [33, 34]. Klotho may have direct phosphaturic effects. In a previous study, we showed that klotho-deficient mice (Kl/Kl) showed upregulation of the intestinal Npt2b mRNA and protein levels, regardless of high plasma FGF23 levels [35]. Dermaku-Sopjani et al. [36] demonstrated that Klotho downregulates the intestinal NaPi-IIb transporter in vitro. Klotho transcript was detectable in murine intestinal tissue. These reports suggest that klotho is essential for FGF23-dependent Npt2b downregulation. However, it is unclear whether klotho simply enhances the FGF23 effect or functions as a direct inhibitor for intestinal NaPi-IIb.

Conclusions

In this review, we have described the roles of $1,25(OH)_2D_3$ in the regulation of renal type II NaPi cotransporters in the kidney and small intestine (fig. 3). Regarding the roles of vitamin D in renal Pi reabsorption, there is a difference in the regulation of the renal NaPi-II transporters between adult and weaning animals. In suckling and weaning animals, the $1,25(OH)_2D_3$/VDR complex is essential for proper development of NaPi-IIa and NaPi-IIc transporter expression in the kidney, while in adult animals, the $1,25(OH)_2D_3$/VDR complex is essential for proper expression of intestinal Npt2b transporters. In addition, $1,25(OH)_2D_3$ stimulates the production of FGF23 in bone and may have a phosphaturic action. $1,25(OH)_2D_3$ also upregulates klotho in the kidney. Klotho leads to reduction of renal NaPi-IIa transporter. Thus, these data suggest that the regulation of renal and intestinal NaPi-II by $1,25(OH)_2D_3$ has a direct action via gene transcription and an indirect action through the phosphaturic function of FGF23/klotho or PTH. However, further studies are needed to explore the possibility that NaPi-II could be a therapeutic target to reduce altered phosphate metabolism and related complications, and to determine whether and how FGF23/klotho might influence the regulation and function of a related protein, NaPi-IIa and NaPi-IIc, which is also involved in phosphate metabolism.

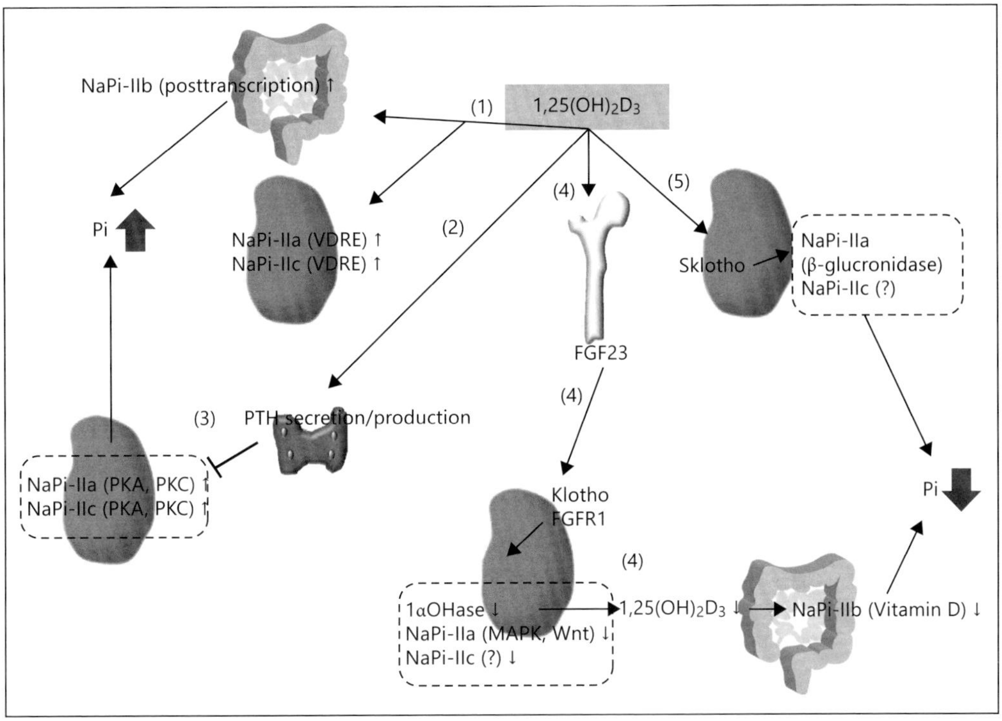

Fig. 3. Action of 1,25(OH)$_2$D$_3$. 1,25(OH)$_2$D$_3$ raises blood Pi concentration by exerting an action through the kidney, intestinal tract and parathyroid gland. At first 1,25(OH)$_2$D$_3$ promotes Pi absorption by posttranscriptionally stimulating the up-regulation of intestinal Npt2b protein (step 1). In addition, 1,25(OH)$_2$D$_3$ promotes the gene transcription of Npt2a and Npt2c in the kidney (step 2). Furthermore, 1,25(OH)$_2$D$_3$ increases Npt2a and Npt2c protein levels by reducing secretion of PTH (step 3). 1,25(OH)$_2$D$_3$ promotes phosphaturic substances. 1,25(OH)$_2$D$_3$ promotes the production of FGF23 (step 4). FGF23 acts on the kidney and reduces the levels of Npt2a and Npt2c protein. FGF23 decreases expression of NaPi-IIb in the small intestine by reducing synthesis of 1,25(OH)$_2$D$_3$. 1,25(OH)$_2$D$_3$ also promotes the production of klotho in the kidney (step 5). Klotho is a phosphaturic factor and modulates the N-glycosylation site of NaPi-IIa. Thus, 1,25(OH)$_2$D$_3$ can produce phosphaturic factors while concomitantly raising blood Pi concentration. VDRE = Vitamin D receptor-binding element; PKA = protein kinase A; PKC = protein kinase C; MAPK = the mitogen-activated protein kinase; Wnt = Wnt/b-catenin signaling.

References

1 Murer H, Hernando N, Forster I, Biber J: Proximal tubular phosphate reabsorption: molecular mechanisms. Physiol Rev 2000:80; 1373–1409.

2 Virkki LV, Biber J, Murer H, Forster IC: Phosphate transporters: a tale of two solute carrier families. Am J Physiol Renal Physiol 2007;293:F643–F654.

3 Miyamoto K, Haito-Sugino S, Kuwahara S, Ohi A, Nomura K, Ito M, Kuwahata M, Kido S, Tatsumi S, Kaneko I, Segawa H: Sodium-dependent phosphate cotransporters: lessons from gene knockout and mutation studies. J Pharm Sci 2011;100:3719–3730.

4 Beck L, Karaplis AC, Amizuka N, Hewson AS, Ozawa H, Tenenhouse HS: Targeted inactivation of Npt2 in mice leads to severe renal phosphate wasting, hypercalciuria, and skeletal abnormalities. Proc Natl Acad Sci USA 1998;95:5372–5377.
5 Segawa H, Onitsuka A, Kuwahata M, Hanabusa E, Furutani J, Kaneko I, Tomoe Y, Aranami F, Matsumoto N, Ito M, Matsumoto M, Li M, Amizuka N, Miyamoto K: Type IIc sodium-dependent phosphate transporter regulates calcium metabolism. J Am Soc Nephrol 2009;20:104–113.
6 Segawa H, Onitsuka A, Furutani J, Kaneko I, Aranami F, Matsumoto N, Tomoe Y, Kuwahata M, Ito M, Matsumoto M, Li M, Amizuka N, Miyamoto K: Npt2a and Npt2c in mice play distinct and synergistic roles in inorganic phosphate metabolism and skeletal development. Am J Physiol Renal Physiol 2009; 297:F671–F678.
7 Sabbagh Y, O'Brien SP, Song W, Boulanger JH, Stokmann A, Arbeeny C, Schiavi SC: Intestinal Npt2b plays a major role in phosphate absorption and homeostasis. J Am Soc Nephrol 2009;20:2348–2358.
8 Haussler MR, Haussler CA, Bartik L, Whitfield GK, Hsieh JC, Slater S, Jurutka PW: Vitamin D receptor: molecular signaling and actions of nutritional ligands in disease prevention. Nutr Rev 2008;66:S98–S112.
9 Razzaque MS: The FGF23-Klotho axis: endocrine regulation of phosphate homeostasis. Nat Rev Endocrinol 2009;5:611–619.
10 Osuka S, Razzaque MS: Can features of phosphate toxicity appear in normophosphatemia? J Bone Miner Metab 2012;30:10–18.
11 Urakawa I, Yamazaki Y, Shimada T, Iijima K, Hasegawa H, Okawa K, Fujita T, Fukumoto S, Yamashita T: Klotho converts canonical FGF receptor into a specific receptor for FGF23. Nature 2006;444:770–774.
12 Kurose H, Ogawa Y, Miyoshi M, Yamamoto M, Nandi A, Rosenblatt KP, Baum MG, Schiavi S, Hu MC, Moe OW, Kuro-o M: Regulation of fibroblast growth factor-23 signaling by klotho. J Biol Chem 2006;281:6120–6123.
13 Saito H, Kusano K, Kinosaki M, Ito H, Hirata M, Segawa H, Miyamoto K, Fukushima N: Human fibroblast growth factor-23 mutants suppress Na^+-dependent phosphate co-transport activity and 1alpha-25-dihydroxyvitamin D_3 production. J Biol Chem 2003;278: 2206–2211.
14 Goldsmith DJ, Cunningham J: Mineral metabolism and vitamin D in chronic kidney disease – more questions than answers. Nat Rev Nephrol 2011:7:341–346.
15 Taketani Y, Segawa H, Chikamori M, Morita K, Tanaka K, Kido S, Yamamoto H, Iemori Y, Tatsumi S, Tsugawa N, Okano T, Kobayashi T, Miyamoto K, Takeda E: Regulation of type II renal Na^+-dependent inorganic phosphate transporters by 1,25-dihydroxyvitamin D_3. Identification of a vitamin D-responsive element in the human NAPi-3 gene. J Biol Chem 1998;273:14575–14581.
16 Friedlaender MM, Wald H, Dranitzki-Elhalel M, Zajicek HK, Levi M, Popovtzer MM: Vitamin D reduces renal NaPi-2 in PTH-infused rats: complexity of vitamin D action on renal P(i) handling. Am J Physiol Renal Physiol 2001;281:F428–F433.
17 Rubinger D, Wald H, Popovtzer MM: 25-Hydroxycholcalciferol amd 1,25-dihydroxycholecalciferol enhance phosphaturia in rats with reduced renal mass: evidence for a PTH-dependent mechanism. Miner Electr Metab 1990;16:348–354.
18 Kurnik BR, Hruska KA: Mechanism of stimulation of renal phosphate transport by 1,25-dihydroxycholecalciferol. Biochim Biophys Acta 1985:817:42–50.
19 Kurnik BR, Huskey M, Hruska KA: 1,25-Dihydroxycholecalciferol stimulates renal phosphate transport by directly altering membrane phosphatidylcholine composition. Biochim Biophys Acta 1987;917:81–85.
20 Masuda M, Yamamoto H, Kozai M, Tanaka S, Ishiguro M, Takei Y, Nakahashi O, Ikeda S, Uebanso T, Taketani Y, Segawa H, Miyamoto K, Takeda E: Regulation of renal sodium-dependent phosphate co-transporter genes (Npt2a and Npt2c) by all-trans-retinoic acid and its receptors. Biochem J 2010;429:583–592.
21 Segawa H, Kaneko I, Yamanaka S, Ito M, Kuwahata M, Inoue Y, Kato S, Miyamoto K: Intestinal Na-P(i) cotransporter adaptation to dietary P(i) content in vitamin D receptor null mice. Am J Physiol Renal Physiol 2004; 287:F39–F47.
22 Kaneko I, Segawa H, Furutani J, Kuwahara S, Aranami F, Hanabusa E, Tominaga R, Giral H, Caldas Y, Levi M, Kato S, Miyamoto K: Hypophosphatemia in vitamin D receptor null mice: effect of rescue diet on the developmental changes in renal Na+ -dependent phosphate cotransporters. Pflügers Arch 2011;461:77–90.

23 Saito H, Maeda A, Ohtomo S, Hirata M, Kusano K, Kato S, Ogata E, Segawa H, Miyamoto K, Fukushima N: Circulating FGF-23 is regulated by 1alpha,25-dihydroxyvitamin D_3 and phosphorus in vivo. J Biol Chem. 2005; 280:2543–2549.
24 Segawa H, Kawakami E, Kaneko I, Kuwahata M, Ito M, Kusano K, Saito H, Fukushima N, Miyamoto K: Effect of hydrolysis-resistant FGF23-R179Q on dietary phosphate regulation of the renal type-II Na/Pi transporter. Pflügers Arch 2003;446:585–592.
25 Inoue Y, Segawa H, Kaneko I, Yamanaka S, Kusano K, Kawakami E, Furutani J, Ito M, Kuwahata M, Saito H, Fukushima N, Kato S, Kanayama HO, Miyamoto K: Role of the vitamin D receptor in FGF23 action on phosphate metabolism. Biochem J 2005;390:325–331.
26 Ohi A, Hanabusa E, Ueda O, Segawa H, Horiba N, Kaneko I, Kuwahara S, Mukai T, Sasaki S, Tominaga R, Furutani J, Aranami F, Ohtomo S, Oikawa Y, Kawase Y, Wada NA, Tachibe T, Kakefuda M, Tateishi H, Matsumoto K, Tatsumi S, Kido S, Fukushima N, Jishage K, Miyamoto K: Inorganic phosphate homeostasis in sodium-dependent phosphate cotransporter Npt2b+/– mice. Am J Physiol Renal Physiol 2011;301:F1105–F1113.
27 Katai K, Miyamoto K, Kishida S, Segawa H, Nii T, Tanaka H, Tani Y, Arai H, Tatsumi S, Morita K, Taketani Y, Takeda E: Regulation of intestinal Na^+-dependent phosphate co-transporters by a low-phosphate diet and 1,25-dihydroxyvitamin D_3. Biochem J 1999; 343:705–712.
28 Segawa H, Kaneko I, Yamanaka S, Ito M, Kuwahata M, Inoue Y, Kato S, Miyamoto K: Intestinal Na-P(i) cotransporter adaptation to dietary P(i) content in vitamin D receptor null mice. Am J Physiol Renal Physiol 2004; 287:F39–F47.
29 Miyamoto K, Ito M, Kuwahata M, Kato S, Segawa H: Inhibition of intestinal sodium-dependent inorganic phosphate transport by fibroblast growth factor 23. Ther Apher Dial 2005;9:331–335.
30 Tomoe Y, Segawa H, Shiozawa K, Kaneko I, Tominaga R, Hanabusa E, Aranami F, Furutani J, Kuwahara S, Tatsumi S, Matsumoto M, Ito M, Miyamoto K: Phosphaturic action of fibroblast growth factor 23 in Npt2 null mice. Am J Physiol Renal Physiol 2010; 298:F1341–F1350.
31 Forster RE, Jurutka PW, Hsieh JC, Haussler CA, Lowmiller CL, Kaneko I, Haussler MR, Whitfield GK: Vitamin D receptor controls expression of the anti-aging klotho gene in mouse and human renal cells. Biochem Biophys Res Commun 2011;414:557–562.
32 Kuro-o M: Klotho and aging. Biochem Biophys Acta 2009;1790:1049–1058.
33 Hu MC, Shi M, Zhang J, Pastor J, Nakatani T, Lanske B, Razzaque S, Rosenblatt KP, Baum MG, Kuro-o M, Moe OW: Klotho: a novel phosphaturic substance acting as an autocrine enzyme in the renal proximal tubule. FASEB J 2010;24:3438–3450.
34 Haung CL, Moe OW: Klotho: a novel regulator of calcium and phosphorus homeostasis. Pflügers Arch Eur J Physiol 2011;462:185–193.
35 Segawa H, Yamanaka S, Ohno Y, Onitsuka A, Shiozawa K, Aranami F, Furutani J, Tomoe Y, Ito M, Kuwahata M, Imura A, Nabeshima Y, Miyamoto K: Correlation between hyperphosphatemia and type II Na-Pi cotransporter activity in klotho mice. Am J Physiol Renal Physiol 2007;292:F769–F779.
36 Dermaku-Sopjani M, Sopjani M, Saxena A, Shojaiefard M, Bogatikov E, Alesutan I, Eichenmuller M, Lang F: Downregulation of NaPi-IIa and NaPi-IIb Na^+-coupled phosphate transporters by coexpression of klotho. Cell Phydiol Biochem 2011;28:251–258.

Ken-ichi Miyamoto, PhD
Department of Molecular Nutrition, Institution of Health Biosciences
The University of Tokushima, Graduate School
3-18-15 Kuramoto-cho, Tokushima 770-8503 (Japan)
E-Mail miyamoto@nutr.med.tokushima-u.ac.jp

Razzaque MS (ed): Phosphate and Vitamin D in Chronic Kidney Disease.
Contrib Nephrol. Basel, Karger, 2013, vol 180, pp 98–109 (DOI: 10.1159/000346789)

Vitamin D in Chronic Kidney Disease

Yan Chun Li

Department of Medicine, The University of Chicago, Chicago, Ill., USA

Abstract

Vitamin D deficiency is highly prevalent in patients with chronic kidney disease (CKD). The low vitamin D status is, to a large extent, caused by dysregulation of vitamin D metabolism as a result of renal insufficiency. Recent studies indicate that vitamin D-deficiency may promote or accelerate the progression of CKD, whereas treatment with low calcemic vitamin D analogs can reduce proteinuria and ameliorate renal damage in animal models of kidney disease and in patients with CKD. The renoprotective activity of vitamin D regulates multiple signaling pathways known to play important roles in renal injury. These findings underscore the importance of correcting vitamin D deficiency with vitamin D supplementation or with activated vitamin D analogs in the management of CKD.

Vitamin D deficiency (usually defined as serum 25-hydroxyvitamin D levels (25(OH)D) <20 ng/ml or 50 nmol/l) is an emerging global health problem that is estimated to affect more than one billion people worldwide [1]. The prevalence of vitamin D deficiency is particularly high in chronic kidney disease (CKD), mostly due to renal insufficiency. The vitamin D endocrine system, with 1,25-dihydroxyvitamin D_3 ($1,25(OH)_2D_3$) as the most active hormonal metabolite, is a pleiotropic hormone system, and its broad biological activities are far beyond the classic regulation of calcium and phosphorus homeostasis. As such, vitamin D deficiency has a detrimental impact not only on the skeleton but also on extraskeletal tissues, including the kidney [2]. Vitamin D therapy has shown potent renal protective effects in animal models as well as in patients with CKD. The goal of this article is to review the current understanding of the renoprotective role of vitamin D in the context of CKD.

High Prevalence of Vitamin D Deficiency in Chronic Kidney Disease

High prevalence of vitamin D deficiency or insufficiency in patients with CKD has been well documented in epidemiological studies published in recent years. Low serum 25(OH)D and $1,25(OH)_2D_3$ levels are very common in patients with CKD even at early stages, and the severity of deficiency increases with the progression of kidney disease [3, 4]. Cross-sectional analyses showed that 25(OH) D deficiency is associated with impaired glomerular filtration rate (GFR) in CKD [5], and 25(OH)D and $1,25(OH)_2D_3$ deficiency are independently associated with decreased hemoglobin levels and anemia in early CKD [6]. Vitamin D deficiency is closely correlated to kidney disease progression, and the serum vitamin D status is an independent inverse predictor of disease progression and death in CKD patients [7], suggesting that vitamin D deficiency may accelerate the progression of CKD. Whether vitamin D deficiency is in fact a cause of CKD progression, however, remains to be determined.

Dysregulation of Vitamin D Metabolism in Chronic Kidney Disease

A number of recent studies have provided evidence that dysregulation of vitamin D metabolism is a main cause for vitamin D deficiency in CKD. While diet is a source of vitamin D, the majority of vitamin D content in the body comes from de novo synthesis in the skin, catalyzed by ultraviolet light irradiation from the sunlight (fig. 1). Once synthesized, vitamin D is quickly converted in the liver to 25(OH)D, the predominant circulating metabolite that commonly used as an indicator of body's vitamin D status. In the circulation 25(OH)D complexes with vitamin D-binding protein (DBP), and the complex is subject to glomerular filtration. After glomerular filtration the 25(OH)D-DBP complex is reabsorbed via megalin-mediated endocytosis in the proximal tubular cells [8], in which 25(OH)D is converted to $1,25(OH)_2D_3$ by renal 25-hydroxyvitamin D 1α-hydroxylase (CYP27B1), the rate-limiting enzyme for $1,25(OH)_2D_3$ production [1]. In renal insufficiency, the decline in megalin-mediated endocytotic activity and renal 1α-hydroxylase activity and the loss of 25(OH)D-DBP into the urine because of proteinuria all contribute to the development of 25(OH)D and $1,25(OH)_2D_3$ deficiency. This notion is supported by recent experimental data that demonstrate dysregulated vitamin D metabolism in kidney disease. Recent studies reported that urinary excretion of DBP is elevated in type 1 diabetes patients [9], and the levels of megalin, DBP and 25(OH)D are increased in the urine of type 1 diabetic mice [10]. In Zucker diabetic fatty rats, the expression of megalin and Dab2 is decreased in the kid-

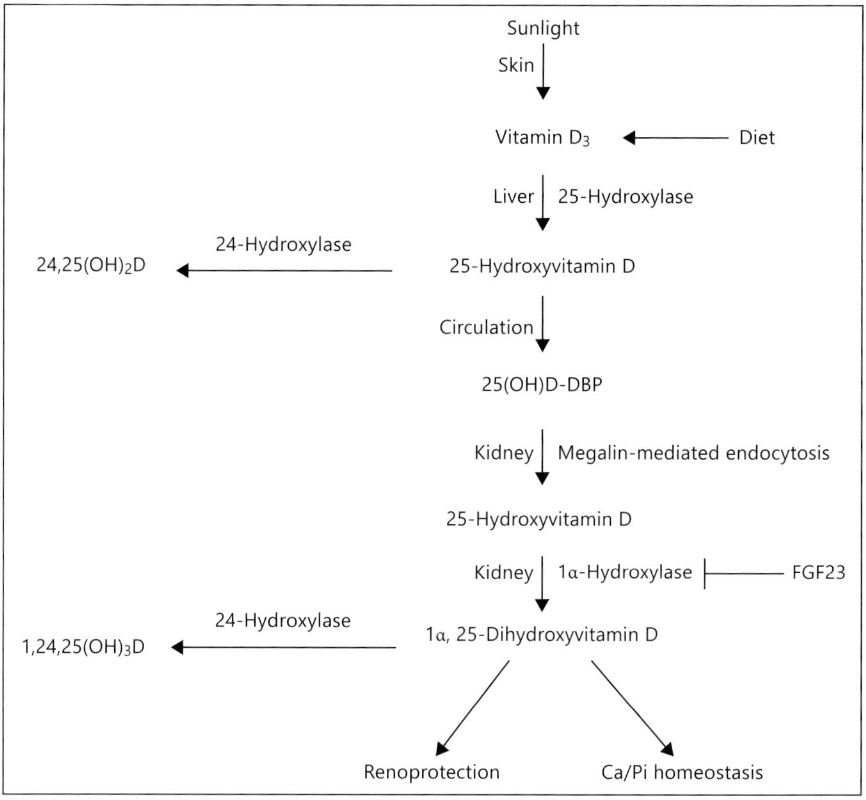

Fig. 1. Vitamin D metabolism in CKD. In renal insufficiency, the megalin-mediated endocytotic activity and renal 1α-hydroxylase activity are declined, and because of proteinuria, the loss of 25(OH)D-DBP into the urine is increased. The increase in FGF23 suppresses 1α-hydroxylase and stimulates 24-hydroxylase. These changes contribute to the development of 25(OH)D and $1,25(OH)_2D_3$ deficiency in CKD. 25(OH)D = 25-hydroxyvitamin D; DBP = vitamin D binding protein; FGF23 = fibroblast growth factor 23.

ney and renal reuptake of 25(OH)D-DBP is compromised, which is accompanied by reduction in 25(OH)D and $1,25(OH)_2D_3$ concentrations in the serum and elevation in urinary excretion of 25(OH)D, $1,25(OH)_2D_3$ and DBP [11]. Moreover, renal 24-hydroxylase (CYP24A1), the main catabolic enzyme involved in 25(OH)D and $1,25(OH)_2D_3$ degradation (fig. 1), was found markedly elevated in uremic rats as well as in diseased human kidney biopsies, together with reduction in serum $1,25(OH)_2D_3$ [12]. These observations suggest that increased CYP24A1 activity also contributes to vitamin D deficiency in kidney disease.

Another key factor that helps to induce low vitamin D status in CKD is fibroblast growth factor (FGF) 23, a major phosphatonin that promotes renal phosphate excretion and suppresses $1,25(OH)_2D_3$ production via inhibition of CYP27B1 and stimulation of CYP24A1 [13] (fig. 1). Circulating FGF23 levels are elevated along with the decline of GFR and the increase in serum creatinine and phosphate levels [14–16]. In advanced CKD, persistent phosphate retention due to impaired kidney function greatly induces FGF23 leading to suppression of $1,25(OH)_2D_3$ production. In fact, FGF23 is also a risk factor for the progression of CKD [17]. In addition, conventional risk factors such as reduced sunlight exposure and poor dietary vitamin D intake also contribute to vitamin D insufficiency in the CKD population (fig. 1).

Effects of Vitamin D Therapy on Mortality in Chronic Kidney Disease

Secondary hyperparathyroidism is a major problem associated with CKD. Chronically elevated parathyroid hormone (PTH) in the patients causes bone loss and fracture. Active vitamin D, in the form $1,25(OH)_2D_3$ or vitamin D analogs, is commonly used to treat the secondary hyperparathyroidism and associated calcium and phosphate abnormalities in CKD patients. In the US, two low-calcemic vitamin D analogs, 1,25-dihydroxy-19-nor-vitamin D_2 (paricalcitol) and 1α-hydroxyvitamin D_2 (doxercalciferol), are approved for this purpose. Interestingly, therapies with vitamin D or its analogs have shown benefits more than the regulation of PTH and mineral homeostasis. A series of retrospective studies have consistently demonstrated significant survival advantage in hemodialysis and nondialysis CKD patients receiving vitamin D or its analog therapy [18–22]. These clinical data suggest that the association between vitamin D therapy and reduced mortality in CKD patients is independent of its effect on PTH and mineral metabolism. On the other hand, data from the Third National Health and Nutrition Examination Survey (NHANES III) in the United States have revealed a correlation between low serum 25(OH) D (<15–17.8 ng/ml) and the risk of all-cause mortality in the general population [23] as well as in subjects with CKD who are not yet undergoing dialysis [24]. Moreover, the Ludwigshafen Risk and Cardiovascular (LURIC) Health Study also demonstrated an association between low vitamin D status and all-cause mortality and cardiovascular mortality in CKD patients in Germany [25]. A meta-analysis of 10 prospective studies confirmed that high 25(OH)D levels are associated with significantly improved survival in patients with CKD [25]. Collectively, these observations warrant large randomized controlled clinical trials to assess whether raising vitamin D levels or correcting vitamin D deficiency indeed ameliorates CKD and reduces mortality in CKD patient population.

Potent Antiproteinuric Activity of Vitamin D in Chronic Kidney Disease

One important renoprotective activity of vitamin D is to reduce albuminuria. Albuminuria is a major risk factor for CKD progression, renal failure, cardiovascular events and death [26], thus reduction of albuminuria is a crucial therapeutic target for CKD management. A cross-sectional analysis of the NHANES III database has revealed a correlation between vitamin D-insufficiency and increased prevalence of albuminuria in the US adult population [27], suggesting that vitamin D has an intrinsic antiproteinuric activity. A number of clinical studies have demonstrated therapeutic efficacy of vitamin D analogs to reduce proteinuria. Following an early report that activated vitamin D analog paricalcitol has significant antiproteinuric activity in CKD [28], two small randomized pilot trials were conducted to evaluate the antiproteinuric effect of oral paricalcitol treatment in CKD patients (n = 24 and n = 61, respectively) [29, 30]. These trials showed that paricalcitol significantly reduced albuminuria independent of its effect on hemodynamics or PTH suppression. A more recent large randomized placebo-controlled clinical trial (the VITAL Study, n = 281) confirmed that paricalcitol is able to reduce albuminuria and blood pressure in patients with diabetic nephropathy who were already on renin-angiotensin inhibitor therapy [31]. This trial showed that at the dose of 2 μg, paricalcitol significantly reduced urinary albumin in the patients by –18 to –28% ($p = 0.014$ vs. placebo). Together these studies have provided strong clinical evidence that low-calcemic vitamin D analogs can be used as an alternative or complementary therapy for treatment of proteinuria in CKD. One limitation, however, in the use of the current approved vitamin D analogs is a relatively narrow therapeutic window. Thus, it is worthwhile to direct efforts towards developing new analogs with less calcemic side effect and better therapeutic efficacy for the management of CKD.

Complex Renoprotective Mechanisms of Vitamin D

Studies with different experimental models of kidney disease have provided insights into the renoprotective mechanism for vitamin D and its analogs. The renoprotective activity of vitamin D is mediated by the vitamin D receptor (VDR), which is abundantly expressed or highly inducible in almost all cell types in the kidney. Research evidence shows that vitamin D protects the kidney by regulating multiple pathways known to play a role in kidney injury, including the renin-angiotensin system (RAS), NF-κB, Wnt/β-catenin and slit diaphragm proteins (fig. 2).

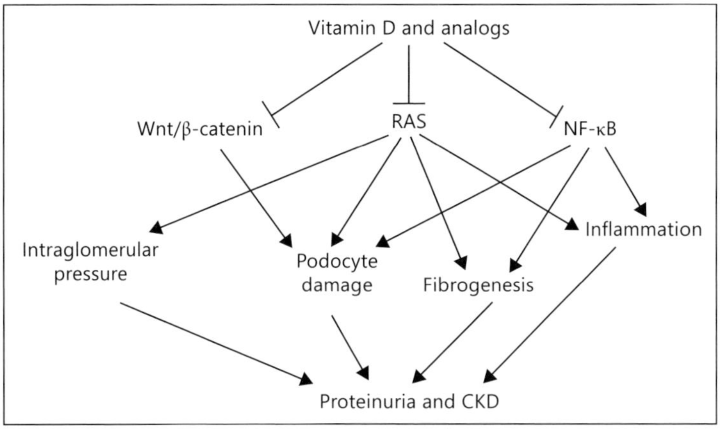

Fig. 2. Renoprotectective mechanisms of vitamin D. Vitamin D and its analogs suppress the RAS, NF-κB activation and Wnt/βcatenin pathway. These pathways play important roles in the development of proteinuria and renal injury. RAS = renin-angiotensin system.

The Renin-Angiotensin System

The intrarenal RAS is a major mediator of renal damage. Many pathological conditions such as hyperglycemia, renal insufficiency and vitamin D deficiency can activate the local RAS in the kidney, leading to increased local concentration of angiotensin (Ang) II, the effector of the RAS cascade that has a broad range of pathogenic activities that promote renal injury [32]. It is well established that the vitamin D hormone is a negative regulator of the RAS by suppressing renin gene expression [33, 34]. VDR-null mutant mice develop more severe renal damage (e.g. increased albuminuria, glomerulosclerosis and interstitial fibrosis) than the wild-type counterparts in diabetic state [35] or under unilateral ureteral obstruction (UUO) [36], as a consequence of enhanced activation of the local RAS in the kidney. In rats with 5/6 nephrectomy paricalcitol treatment attenuated glomerular and tubulointerstitial damage and reduced blood pressure and proteinuria by blocking the activation of the local RAS in the kidney remnant [37]. Vitamin D analog doxercalciferol modulated high fat-induced renal disease by targeting the RAS and lipid metabolism [38]. Moreover, many studies have demonstrated that combination therapy with one vitamin D analog (paricalcitol or doxercalciferol) and one RAS inhibitor (ACE inhibitor or ARB) produced synergistic or additive therapeutic effects in blocking renal damage in experimental models of type 1 and type 2 diabetes mellitus [39–42], UUO [43] and 5/6 nephrectomy [44]. The basis for the enhanced renoprotective efficacy of this combination therapy is the blockade of the compensatory renin induction commonly encountered in the use of RAS inhibitors by the vitamin D analog.

Inhibition of the renin induction and Ang II accumulation within the kidney leads to significantly better therapeutic outcomes [2]. The combination strategy demonstrated in these preclinical studies may explain why vitamin D analogs are still effective in reducing proteinuria in CKD patients who are already on RAS inhibitor therapy [29, 31]. Given that RAS inhibitors are a standard of care for CKD and many other diseases, an important implication of the combination therapy is that vitamin D analogs may be used as a complementary therapy in many human diseases.

NF-κB Activation and Inflammation

NF-κB is a family of transcription factors that function as a master regulator of immune response. NF-κB regulates a wide range of genes involved in inflammation, proliferation and fibrogenesis that are known to play important roles in kidney disease [45]. Vitamin D suppresses NF-κB activation, partly by blocking its nuclear translocation and disrupting its DNA binding [46]. Serum vitamin D status is inversely associated with increased renal inflammation in patients with a variety of kidney disorders [47]. Recent investigations showed that paricalcitol protects against kidney damage by blocking NF-κB activity and reducing renal inflammation in experimental models of kidney disease such as UUO [48], cyclosporine-induced nephropathy [49] and gentamicin-induced renal injury [50]. At the molecular level, $1,25(OH)_2D_3$ inhibits the expression of high glucose-induced MCP-1 and angiotensinogen and inflammation-induced PAI-1 by blocking the activation of NF-κB [51–53]. These factors are involved in renal disease progression. For example, MCP-1 induces macrophage infiltration in the kidney, where activated macrophages release factors and mediators that promote kidney disease progression. It is known that Ang II activates NF-κB, whereas NF-κB mediates high glucose induction of angiotensinogen, the substrate of renin for Ang II production. This vicious circle, which enhances the local accumulation of Ang II in diabetic nephropathy, can be disrupted by vitamin D [2].

Wnt/β-Catenin Pathway

Recent studies have implicated that activation of the Wnt/β-catenin pathway induces podocyte injury and albuminuria in adriamycin nephropathy [54]. $1,25(OH)_2D_3$ can block the Wnt/β-catenin pathway by promoting the physical interaction between VDR and β-catenin, thus preventing β-catenin nuclear translocation, and the molecular basis underlying liganded VDR and β-catenin interaction has been established [55]. Indeed, paricalcitol can prevent podocyte dysfunction and ameliorate proteinuria by blocking Wnt/β-catenin signaling via VDR-β-catenin interaction [56].

Slit Diaphragm Proteins

In the glomerular filtration barrier the slit diaphragm functions as the size-selective and charge-selective barrier to protein leakage. A number of proteins involved in the formation of the slit diaphragm are downregulated in diabetes, and it has been reported that vitamin D therapy reverses this downregulation [39, 41]. This vitamin D effect on the slit diaphragm proteins may result from a direct stimulation on these genes, or is secondary to the prevention of podocyte loss, as most of these proteins are produced by podocytes. Given the importance of podocytes in the regulation of glomerular filtration, it is speculated that podocytes are a key antiproteinuric target of vitamin D [2]. Indeed, recent transgenic studies have demonstrated that podocyte VDR signaling prevents podocyte apoptosis and attenuates albuminuria in diabetic nephropathy models [57]. Furthermore, nephrin, a slit diaphragm protein produced by podocytes that plays a central role in the integrity of the slit diaphragm [58], is a direct upregulatory target of vitamin D. Nephrin is highly inducible by $1,25(OH)_2D_3$ in podocyte cultures [59], and a *cis*-vitamin D response element has been identified in the proximal promoter of the nephrin gene that mediates its upregulation by $1,25(OH)_2D_3$ [60]. Therefore, the vitamin D-VDR signaling plays a critical role to maintain the integrity of the slit diaphragm and protect the glomerular filtration barrier, which contributes to the antiproteinuric effect of vitamin D.

Conclusion

The incident of CKD is persistently rising around the globe, which calls for new and more effective therapeutic methods for this devastating disease. Growing evidence suggest that vitamin D-deficiency in CKD not only affects the skeleton and mineral homeostasis, but may also accelerate renal disease progression. Therapies with vitamin D and its analogs have been proved in principle to be effective in reducing proteinuria and preventing kidney disease progression. Therefore, the advance in the understanding of the renoprotective function of the vitamin D endocrine system and its interactions with FGF23/Klotho system [61, 62] may guide new therapeutic strategies for the treatment and management of patients suffering from various stages of CKD.

References

1 Holick MF: Vitamin D deficiency. N Engl J Med 2007;357:266–281.

2 Li YC: Renoprotective effects of vitamin D analogs. Kidney Int 2010;78:134–139.

3 Levin A, Bakris GL, Molitch M, Smulders M, Tian J, Williams LA, Andress DL: Prevalence of abnormal serum vitamin D, PTH, calcium, and phosphorus in patients with chronic kidney disease: results of the study to evaluate early kidney disease. Kidney Int 2007;71:31–38.

4 Levin A, Le Barbier M, Er L, Andress D, Sigrist MK, Djurdjev O: Incident isolated 1,25(OH)2D3 deficiency is more common than 25(OH)D deficiency in CKD. J Nephrol 2012;25:204–210.

5 Urena-Torres P, Metzger M, Haymann JP, Karras A, Boffa JJ, Flamant M, Vrtovsnik F, Gauci C, Froissart M, Houillier P, Stengel B: Association of kidney function, vitamin D deficiency, and circulating markers of mineral and bone disorders in CKD. Am J Kidney Dis 2011;58:544–553.

6 Patel NM, Gutierrez OM, Andress DL, Coyne DW, Levin A, Wolf M: Vitamin D deficiency and anemia in early chronic kidney disease. Kidney Int 2010;77:715–720.

7 Ravani P, Malberti F, Tripepi G, Pecchini P, Cutrupi S, Pizzini P, Mallamaci F, Zoccali C: Vitamin D levels and patient outcome in chronic kidney disease. Kidney Int 2009;75: 88–95.

8 Nykjaer A, Dragun D, Walther D, Vorum H, Jacobsen C, Herz J, Melsen F, Christensen EI, Willnow TE: An endocytic pathway essential for renal uptake and activation of the steroid 25-(OH) vitamin D_3. Cell 1999;96: 507–515.

9 Thrailkill KM, Jo CH, Cockrell GE, Moreau CS, Fowlkes JL: Enhanced excretion of vitamin D binding protein in type 1 diabetes: a role in vitamin D deficiency? J Clin Endocrinol Metab 2011;96:142–149.

10 Fowlkes JL, Bunn RC, Cockrell GE, Clark LM, Wahl EC, Lumpkin CK, Thrailkill KM: Dysregulation of the intrarenal vitamin D endocytic pathway in a nephropathy-prone mouse model of type 1 diabetes. Exp Diab Res 2011;Article ID 269378.

11 Anderson RL, Ternes SB, Strand KA, Rowling MJ: Vitamin D homeostasis is compromised due to increased urinary excretion of the 25-hydroxycholecalciferol-vitamin D-binding protein complex in the Zucker diabetic fatty rat. Am J Physiol Endocrinol Metab 2010;299:E959–E967.

12 Helvig CF, Cuerrier D, Hosfield CM, Ireland B, Kharebov AZ, Kim JW, Ramjit NJ, Ryder K, Tabash SP, Herzenberg AM, Epps TM, Petkovich M: Dysregulation of renal vitamin D metabolism in the uremic rat. Kidney Int 2010;78:463–472.

13 Shimada T, Hasegawa H, Yamazaki Y, Muto T, Hino R, Takeuchi Y, Fujita T, Nakahara K, Fukumoto S, Yamashita T: FGF-23 is a potent regulator of vitamin D metabolism and phosphate homeostasis. J Bone Miner Res 2004;19:429–435.

14 Imanishi Y, Inaba M, Nakatsuka K, Nagasue K, Okuno S, Yoshihara A, Miura M, Miyauchi A, Kobayashi K, Miki T, Shoji T, Ishimura E, Nishizawa Y: FGF-23 in patients with end-stage renal disease on hemodialysis. Kidney Int 2004;65:1943–1946.

15 Larsson T, Nisbeth U, Ljunggren O, Juppner H, Jonsson KB: Circulating concentration of FGF-23 increases as renal function declines in patients with chronic kidney disease, but does not change in response to variation in phosphate intake in healthy volunteers. Kidney Int 2003;64:2272–2279.

16 Shigematsu T, Kazama JJ, Yamashita T, Fukumoto S, Hosoya T, Gejyo F, Fukagawa M: Possible involvement of circulating fibroblast growth factor 23 in the development of secondary hyperparathyroidism associated with renal insufficiency. Am J Kidney Dis 2004;44: 250–256.

17 Wolf M: Update on fibroblast growth factor 23 in chronic kidney disease. Kidney Int 2012;82:737–747.

18 Kovesdy CP: Survival benefits with vitamin D receptor activation: new insights since 2003. Clin J Am Soc Nephrol 2010;5:1704–1709.

19 Teng M, Wolf M, Ofsthun MN, Lazarus JM, Hernan MA, Camargo CA Jr, Thadhani R: Activated injectable vitamin D and hemodialysis survival: a historical cohort study. J Am Soc Nephrol 2005;16:1115–1125.

20 Tentori F, Hunt WC, Stidley CA, Rohrscheib MR, Bedrick EJ, Meyer KB, Johnson HK, Zager PG: Mortality risk among hemodialysis patients receiving different vitamin D analogs. Kidney Int 2006;70:1858–1865.
21 Kovesdy CP, Ahmadzadeh S, Anderson JE, Kalantar-Zadeh K: Association of activated vitamin D treatment and mortality in chronic kidney disease. Arch Intern Med 2008;168: 397–403.
22 Shoben AB, Rudser KD, de Boer IH, Young B, Kestenbaum B: Association of oral calcitriol with improved survival in nondialyzed CKD. J Am Soc Nephrol 2008;19: 1613–1619.
23 Melamed ML, Michos ED, Post W, Astor B: 25-hydroxyvitamin D levels and the risk of mortality in the general population. Arch Intern Med 2008;168:1629–1637.
24 Mehrotra R, Kermah DA, Salusky IB, Wolf MS, Thadhani RI, Chiu YW, Martins D, Adler SG, Norris KC: Chronic kidney disease, hypovitaminosis D, and mortality in the United States. Kidney Int 2009;76:977–983.
25 Pilz S, Iodice S, Zittermann A, Grant WB, Gandini S: Vitamin D status and mortality risk in CKD: a meta-analysis of prospective studies. Am J Kidney Dis 2011;58:374–382.
26 Keane WF, Eknoyan G: Proteinuria, albuminuria, risk, assessment, detection, elimination (PARADE): a position paper of the National Kidney Foundation. Am J Kidney Dis 1999;33:1004–1010.
27 de Boer IH, Ioannou GN, Kestenbaum B, Brunzell JD, Weiss NS: 25-Hydroxyvitamin D levels and albuminuria in the Third National Health and Nutrition Examination Survey (NHANES III). Am J Kidney Dis 2007;50:69–77.
28 Agarwal R, Acharya M, Tian J, Hippensteel RL, Melnick JZ, Qiu P, Williams L, Batlle D: Antiproteinuric effect of oral paricalcitol in chronic kidney disease. Kidney Int 2005;68: 2823–2828.
29 Alborzi P, Patel NA, Peterson C, Bills JE, Bekele DM, Bunaye Z, Light RP, Agarwal R: Paricalcitol reduces albuminuria and inflammation in chronic kidney disease: a randomized double-blind pilot trial. Hypertension 2008;52:249–255.
30 Fishbane S, Chittineni H, Packman M, Dutka P, Ali N, Durie N: Oral paricalcitol in the treatment of patients with CKD and proteinuria: a randomized trial. Am J Kidney Dis 2009;54:647–652.
31 de Zeeuw D, Agarwal R, Amdahl M, Audhya P, Coyne D, Garimella T, Parving HH, Pritchett Y, Remuzzi G, Ritz E, Andress D: Selective vitamin D receptor activation with paricalcitol for reduction of albuminuria in patients with type 2 diabetes (VITAL study): a randomised controlled trial. Lancet 2010; 376:1543–1551.
32 Carey RM, Siragy HM: The intrarenal renin-angiotensin system and diabetic nephropathy. Trends Endocrinol Metab 2003;14:274–281.
33 Li YC, Kong J, Wei M, Chen ZF, Liu SQ, Cao LP: 1,25-Dihydroxyvitamin D(3) is a negative endocrine regulator of the renin-angiotensin system. J Clin Invest 2002;110:229–238.
34 Yuan W, Pan W, Kong J, Zheng W, Szeto FL, Wong KE, Cohen R, Klopot A, Zhang Z, Li YC: 1,25-Dihydroxyvitamin D3 suppresses renin gene transcription by blocking the activity of the cyclic AMP response element in the renin gene promoter. J Biol Chem 2007; 282:29821–29830.
35 Zhang Z, Sun L, Wang Y, Ning G, Minto AW, Kong J, Quigg RJ, Li YC: Renoprotective role of the vitamin D receptor in diabetic nephropathy. Kidney Int 2008;73:163–171.
36 Zhang Y, Kong J, Deb DK, Chang A, Li YC: Vitamin D receptor attenuates renal fibrosis by suppressing the renin-angiotensin system. J Am Soc Nephrol 2010;21:966–973.
37 Freundlich M, Quiroz Y, Zhang Z, Zhang Y, Bravo Y, Weisinger JR, Li YC, Rodriguez-Iturbe B: Suppression of renin-angiotensin gene expression in the kidney by paricalcitol. Kidney Int 2008;74:1394–1402.
38 Wang XX, Jiang T, Shen Y, Santamaria H, Solis N, Arbeeny C, Levi M: Vitamin D receptor agonist doxercalciferol modulates dietary fat-induced renal disease and renal lipid metabolism. Am J Physiol Renal Physiol 2011;300:F801–F810.
39 Zhang Z, Zhang Y, Ning G, Deb DK, Kong J, Li YC: Combination therapy with AT1 blocker and vitamin D analog markedly ameliorates diabetic nephropathy: blockade of compensatory renin increase. Proc Natl Acad Sci USA 2008;105:15896–15901.

40 Deb DK, Sun T, Wong KE, Zhang Z, Ning G, Zhang Y, Kong J, Shi H, Chang A, Li YC: Combined vitamin D analog and AT1 receptor antagonist synergistically block the development of kidney disease in a model of type 2 diabetes. Kidney Int 2010;77:1000–1009.
41 Zhang Y, Deb DK, Kong J, Ning G, Wong Y, Li G, Chen Y, Zhang Z, Strugnell S, Sabbagh Y, Arbeeny C, Li YC: Long-term therapeutic effect of vitamin D analog doxercalciferol on diabetic nephropathy: strong synergism with AT1 receptor antagonist. Am J Physiol Ren Physiol 2009;297:F791–F801.
42 Ohara I, Tanimoto M, Gohda T, Yamazaki T, Hagiwara S, Murakoshi M, Aoki T, Toyoda H, Ishikawa Y, Funabiki K, Horikoshi S, Tomino Y: Effect of combination therapy with angiotensin receptor blocker and 1,25-dihydroxyvitamin D(3) in type 2 diabetic nephropathy in KK-A(y)/Ta mice. Nephron Exp Nephrol 2011;117:e124–e132.
43 Tan X, He W, Liu Y: Combination therapy with paricalcitol and trandolapril reduces renal fibrosis in obstructive nephropathy. Kidney Int 2009;76:1248–1257.
44 Mizobuchi M, Morrissey J, Finch JL, Martin DR, Liapis H, Akizawa T, Slatopolsky E: Combination therapy with an angiotensin-converting enzyme inhibitor and a vitamin D analog suppresses the progression of renal insufficiency in uremic rats. J Am Soc Nephrol 2007;18:1796–1806.
45 Guijarro C, Egido J: Transcription factor-kappa B (NF-kappa B) and renal disease. Kidney Int 2001;59:415–424.
46 Sun J, Kong J, Duan Y, Szeto FL, Liao A, Madara JL, Li YC: Increased NF-{kappa}B activity in fibroblasts lacking the vitamin D receptor. Am J Physiol Endocrinol Metab 2006;291:E315–E322.
47 Zehnder D, Quinkler M, Eardley KS, Bland R, Lepenies J, Hughes SV, Raymond NT, Howie AJ, Cockwell P, Stewart PM, Hewison M: Reduction of the vitamin D hormonal system in kidney disease is associated with increased renal inflammation. Kidney Int 2008;74:1343–1353.
48 Tan X, Wen X, Liu Y: Paricalcitol inhibits renal inflammation by promoting vitamin D receptor-mediated sequestration of NF-kappaB signaling. J Am Soc Nephrol 2008;19:1741–1752.
49 Park JW, Bae EH, Kim IJ, Ma SK, Choi C, Lee J, Kim SW: Paricalcitol attenuates cyclosporine-induced kidney injury in rats. Kidney Int 2010;77:1076–1085.
50 Park JW, Bae EH, Kim IJ, Ma SK, Choi C, Lee J, Kim SW: Renoprotective effects of paricalcitol on gentamicin-induced kidney injury in rats. Am J Physiol Renal Physiol 2010;298:F301–F313.
51 Deb DK, Chen Y, Zhang Z, Zhang Y, Szeto FL, Wong KE, Kong J, Li YC: 1,25-Dihydroxyvitamin D_3 suppresses high glucose-induced angiotensinogen expression in kidney cells by blocking the NF-{kappa}B pathway. Am J Physiol Renal Physiol 2009;296:F1212–F1218.
52 Zhang Z, Yuan W, Sun L, Szeto FL, Wong KE, Li X, Kong J, Li YC: 1,25-Dihydroxyvitamin D(3) targeting of NF-kappaB suppresses high glucose-induced MCP-1 expression in mesangial cells. Kidney Int 2007;72:193–201.
53 Chen Y, Kong J, Sun T, Li G, Szeto FL, Liu W, Deb DK, Wang Y, Zhao Q, Thadhani R, Li YC: 1,25-Dihydroxyvitamin D(3) suppresses inflammation-induced expression of plasminogen activator inhibitor-1 by blocking nuclear factor-kappaB activation. Arch Biochem Biophys 2011;507:241–247.
54 Dai C, Stolz DB, Kiss LP, Monga SP, Holzman LB, Liu Y: Wnt/beta-catenin signaling promotes podocyte dysfunction and albuminuria. J Am Soc Nephrol 2009;20:1997–2008.
55 Shah S, Islam MN, Dakshanamurthy S, Rizvi I, Rao M, Herrell R, Zinser G, Valrance M, Aranda A, Moras D, Norman A, Welsh J, Byers SW: The molecular basis of vitamin D receptor and beta-catenin crossregulation. Mol Cell 2006;21:799–809.
56 He W, Kang YS, Dai C, Liu Y: Blockade of Wnt/beta-catenin signaling by paricalcitol ameliorates proteinuria and kidney injury. J Am Soc Nephrol 2011;22:90–103.
57 Wang Y, Deb DK, Zhang Z, Sun T, Liu W, Yoon D, Kong J, Chen Y, Chang A, Li YC: Renoprotective role of podocyte vitamin D receptor signaling in diabetic nephropathy. J Am Soc Nephrol 2012;23:1977–1986.

58 Ruotsalainen V, Ljungberg P, Wartiovaara J, Lenkkeri U, Kestila M, Jalanko H, Holmberg C, Tryggvason K: Nephrin is specifically located at the slit diaphragm of glomerular podocytes. Proc Natl Acad Sci USA 1999;96: 7962–7967.
59 Takano Y, Yamauchi K, Hiramatsu N, Kasai A, Hayakawa K, Yokouchi M, Yao J, Kitamura M: Recovery and maintenance of nephrin expression in cultured podocytes and identification of HGF as a repressor of nephrin. Am J Physiol Renal Physiol 2007;292:F1573–F1582.
60 Deb DK, Wang Y, Zhang Z, Nie H, Huang X, Yuan Z, Chen Y, Zhao Q, Li YC: Molecular mechanism underlying 1,25-dihydroxyvitamin D regulation of nephrin gene expression. J Biol Chem 2011;286:32011–32017.
61 Razzaque MS: The FGF23-Klotho axis: endocrine regulation of phosphate homeostasis. Nat Rev Endocrinol 2009;5:611–619.
62 Quarles LD: Skeletal secretion of FGF-23 regulates phosphate and vitamin D metabolism. Nat Rev Endocrinol 2012;8:276–286.

Yan Chun Li, PhD
Department of Medicine, The University of Chicago
900 E. 57th Street, KCBD, Mailbox 9-9110
Chicago, IL 60637 (USA)
E-Mail cyan@medicine.bsd.uchicago.edu

Razzaque MS (ed): Phosphate and Vitamin D in Chronic Kidney Disease.
Contrib Nephrol. Basel, Karger, 2013, vol 180, pp 110–123 (DOI: 10.1159/000346791)

Parathyroid Function in Chronic Kidney Disease: Role of FGF23-Klotho Axis

Masahiro Koizumi · Hirotaka Komaba · Masafumi Fukagawa

Division of Nephrology, Endocrinology and Metabolism, Tokai University School of Medicine, Isehara, Japan

Abstract

The parathyroid gland plays a central role in the regulation of mineral metabolism. In patients with chronic kidney disease (CKD), circulating levels of parathyroid hormone (PTH) are progressively increased as kidney function declines, as a result of phosphate retention, hypocalcemia, decreased production of 1,25-dihydroxyvitamin D [1,25$(OH)_2$D], endogenous changes within the parathyroid gland, and skeletal resistance to the actions of PTH. In addition, the identification of fibroblast growth factor 23 (FGF23) and its cofactor Klotho offers important implications for the deeper understanding of disordered mineral metabolism in CKD. In early CKD, increased FGF23 to maintain neutral phosphate balance results in suppression of renal 1,25$(OH)_2$D production and thereby triggers the early development of secondary hyperparathyroidism. FGF23 also acts directly on the parathyroid to decrease PTH synthesis and secretion, but this effect is blunted in advanced stages of CKD, due to decreased expression of the Klotho-FGF receptor 1 complex and increased concentrations of C-terminal FGF23 that competes with full-length FGF23 for binding to the receptor complex. Recent clinical studies also reported that high levels of FGF23 are associated with morbidity and mortality as well as treatment resistance to active vitamin D, suggesting the potential of FGF23 as a novel biomarker to guide treatment of disordered phosphate metabolism in CKD. This review will discuss the pathogenesis of secondary hyperparathyroidism, particularly focusing on the emerging role of the FGF23-Klotho axis in patients with CKD.

The parathyroid gland plays a central role in the regulation of mineral metabolism by secreting parathyroid hormone (PTH) [1]. PTH acts on the kidney to reduce urinary calcium excretion, inhibit phosphate reabsorption, and stimulate the production of 1,25-dihydroxy vitamin D [1,25$(OH)_2$D]. The resulting increased

$1,25(OH)_2D$ acts on the intestines to increase dietary absorption of calcium and phosphate. PTH also acts on the bone to increase the efflux of calcium and phosphate. The secretion of PTH is regulated by serum ionized calcium acting via the parathyroid calcium-sensing receptors (CaSRs) and by $1,25(OH)_2D$ via the vitamin D receptors (VDRs). Another possible regulator of PTH is phosphate, although the sensing mechanisms remain to be identified [2]. In addition, it has recently been shown that fibroblastic growth factor 23 (FGF23), a bone-derived phosphaturic hormone [3], also regulates the synthesis and secretion of PTH [4].

In patients with chronic kidney disease (CKD), circulating levels of PTH are progressively increased as kidney function declines, known as secondary hyperparathyroidism (SHPT) [5, 6]. SHPT is characterized by excessive synthesis of PTH and parathyroid hyperplasia. Severe SHPT results in derangements of mineral and bone metabolism and also contributes to the progression of vascular calcification, leading to increased risk for cardiovascular and all-cause mortality [7]. While it has been well accepted that development of SHPT is caused by phosphate retention, decreased levels of $1,25(OH)_2D$, and resultant hypocalcemia, recent works suggest an important role of FGF23-Klotho axis in the pathogenesis of SHPT [8].

In this review, we will discuss the pathogenesis of SHPT, particularly focusing on the interactions between the parathyroid gland and the FGF23-Klotho axis. The effects of treatment for SHPT on the FGF23-Klotho axis will be also highlighted.

Traditional View of the Pathogenesis of SHPT

Many years of active research have demonstrated that the major contributors of SHPT include phosphate retention, hypocalcemia, decreased production of $1,25(OH)_2D$, endogenous changes within the parathyroid gland, and skeletal resistance to the actions of PTH [9]. These are pathophysiologically related to each other, and one or more of these factors play roles in the genesis of SHPT at different degrees during the course of CKD.

Phosphate Retention

The contribution of phosphate retention to the pathogenesis of SHPT has been determined by a series of studies. The most traditional concept, originally proposed by Bricker et al, was that phosphate retention, as a result of decreased kidney function, would cause transient hypocalcemia, which would, in turn, trigger PTH secretion to maintain mineral homeostasis [10]. This concept was based on their observations that dietary phosphate restriction resulted in amelioration of SHPT [11]. Another group also found that restriction of dietary intake decreased the plasma concentration of $1,25(OH)_2D$ and vice versa, and proposed that phosphate retention contributes indirectly to the development of SHPT via decrease in the levels of $1,25(OH)_2D$ [12]. Experimental studies also showed that high phosphate concentration stimulated PTH synthesis and secretion directly in the absence of changes in ionized calcium in vitro [13]. Although the precise mechanism by which phosphate stimulates PTH secretion remains to be determined, it has been shown that phosphate acts on the PTH gene posttranscriptionally and determines the stability of the PTH mRNA [14].

Decrease in the Levels of $1,25(OH)_2D$

The kidney is the principal site for the production of $1,25(OH)_2D$. In the setting of CKD, decreases in kidney mass lead to progressive reductions in serum levels of $1,25(OH)_2D$. As $1,25(OH)_2D$ decreases PTH secretion by decreasing the transcription of the PTH gene, reduced renal synthesis of $1,25(OH)_2D$ is one of the root causes of the development of SHPT. As mentioned above, phosphate directly inhibits 1α-hydroxylase activity in the kidney and thereby contributes to decreased production of $1,25(OH)_2D$ [12]. Recent studies also showed an additional mechanism of decreased $1,25(OH)_2D$ in patients with CKD. One of the candidates is the decreased delivery of 25-hydroxyvitamin D [25(OH)D] to the proximal tubular cells, where 25(OH)D is converted into $1,25(OH)_2D$ by 1α-hydroxylase [15]. Another is the phosphaturic hormone FGF23, of which role in the pathogenesis of SHPT will be described later.

Endogenous Changes in the Parathyroid Gland

Parathyroid cells are generally quiescent and rarely divide in normal physiological conditions [16], but in the setting of CKD, several mitogenic stimuli such as phosphate retention, hypocalcemia, and decreased $1,25(OH)_2D$ induces hypertrophy and proliferation of parathyroid cells, initially leading to diffuse hyperplasia [5, 6]. In the advanced stage, some cells proliferate vigorously and form small nodules. These nodules then enlarge grossly and become encapsulated, leading to nodular hyperplasia. In the most severe cases, one of these nodules occupies the whole gland (single nodule) [17]. The development and progression of parathyroid hyperplasia is associated with downregulation of VDR and CaSR [18, 19], potentially leading to increased secretion of PTH due to reduced response to $1,25(OH)_2D$ and calcium ions. These endogenous changes may also explain the resistance to active vitamin D therapy in patients with advanced SHPT [20]. Besides the downregulation of VDR and CaSR, recent studies have shown that enhanced parathyroid expression of the potent growth promoter transforming growth factor-α and its receptor, the epidermal growth factor receptor, play a role in the aggressive parathyroid growth seen in CKD [21].

Skeletal Resistance to PTH

The calcemic action of PTH is blunted in patients with uremia, referred to as skeletal resistance to PTH [22]. This phenomenon is another factor that contributes to the development of SHPT. Many factors are have been postulated to be involved in skeletal resistance to PTH, including phosphate retention, decreased levels of 1,25$(OH)_2$D, downregulation of the PTH receptor, accumulation of 7–84PTH fragments that competitively inhibit the action of bioactive 1–84PTH, accumulation of osteoprotegerin, decreased bone morphogenetic protein-7, and presence of uremic toxins. Among the wide variety of uremic toxins, it has been shown that indoxyl sulfate augments oxidative stress, and thereby impairs osteoblast function and downregulates PTH receptor expression on osteoblasts [23]. Conversely, the administration of an oral charcoal adsorbent AST-120, which inhibits the accumulation of indoxyl sulfate, has been shown to increase the action of PTH [24].

Role of FGF23-Klotho Axis in the Pathogenesis of SHPT

The identification and characterization of FGF23 and Klotho have provided important clues toward understanding of the pathogenesis of SHPT [8]. FGF23 is secreted mainly by osteocytes and acts primarily on the kidney to induce urinary phosphate excretion through downregulation of sodium-phosphate cotransporters and suppress renal production of 1,25$(OH)_2$D by inhibiting 1α-hydroxylase and stimulating 24-hydroxylase [3]. These effects are dependent on the presence of the membrane protein Klotho [25], which converts ubiquitous FGF receptor 1 (FGFR1) into a specific receptor for FGF23 [26, 27]. Klotho is expressed in the kidney, parathyroid gland, pituitary gland, and choroid plexus, and determines the tissue specificity of FGF23.

FGF23-Klotho Axis in Early CKD

As described above, it has been considered that the major contributors of SHPT include phosphate retention, decreased 1,25$(OH)_2$D, and resultant hypocalcemia. Among these factors, a progressive decline in 1,25$(OH)_2$D synthesis is likely to be the primary initiating mechanism, which has been attributed to insufficient 1α-hydroxylase activity due to decreased renal mass and phosphate retention. However, this hypothesis cannot explain the clinical observation that serum 1,25$(OH)_2$D begins to decline in early kidney disease long before the development of overt hyperphosphatemia. Recent clinical and experimental data suggest that FGF23 play an important role in the initial decline in 1,25$(OH)_2$D and subsequent hypersecretion of PTH in early-stage CKD.

In patients with early-stage CKD, serum FGF23 increases progressively as kidney function declines [28, 29]. This occurs presumably as an appropriate physiologic response to maintain neutral phosphate balance by promoting urinary phosphate excretion. In one of these studies, FGF23 levels were inversely associated with $1,25(OH)_2D$ levels, independently of key factors known to influence $1,25(OH)_2D$, including serum phosphate, renal function, PTH, and 25(OH)D [29]. This finding suggests that in patients with CKD, increased FGF23 to maintain neutral phosphate balance results in suppression of renal $1,25(OH)_2D$ production and thereby triggers the early development of SHPT. This hypothesis has recently been supported by an experimental study using an animal model of CKD and anti-FGF23 antibodies [30].

FGF23-Klotho Axis in Advanced CKD

In advanced stages of CKD, dietary phosphate may overcome such compensation for decreased nephron mass despite progressively increased FGF23 levels. This results in overt hyperphosphatemia and $1,25(OH)_2D$ deficiency, both of which further stimulate PTH secretion and parathyroid cell proliferation, leading to the development of parathyroid hyperplasia [8]. Although still unclear, the markedly increased levels of FGF23 in this setting would be caused by chronic phosphate retention and active vitamin D therapy [31]. Recent experimental studies also showed that FGF23 secretion in osteocytes is directly regulated by PTH both in vivo and in vitro [32] (fig. 1).

Besides the indirect effect of FGF23 on the development of SHPT by decreasing the renal synthesis of $1,25(OH)_2D$, FGF23 also acts directly on the parathyroid to suppress PTH secretion through the Klotho-FGFR1 complex [4]. However, in patients undergoing dialysis, PTH secretion remains stimulated despite extremely high FGF23 levels [33]. Recent research has shown that the parathyroid expression of Klotho-FGFR1 complex is progressively downregulated in patients with SHPT [34], which likely contributes to the resistance to the inhibitory effect of FGF23 on PTH secretion. Experimental studies in rats with CKD also demonstrated that increased circulating levels of FGF23 fail to decrease PTH secretion as a result of downregulation of this receptor complex [35]. Another possible mechanism of the parathyroid resistance to FGF23 is increased concentrations of C-terminal tail of FGF23 in CKD, which impairs FGF23 signaling by competing with full-length ligand for binding to the Klotho-FGFR1 complex [36]. It is also possible but remains to be proven that some uremic toxins interfere with interaction between FGF23 and Klotho-FGFR1c complex and/or subsequent intracellular signaling.

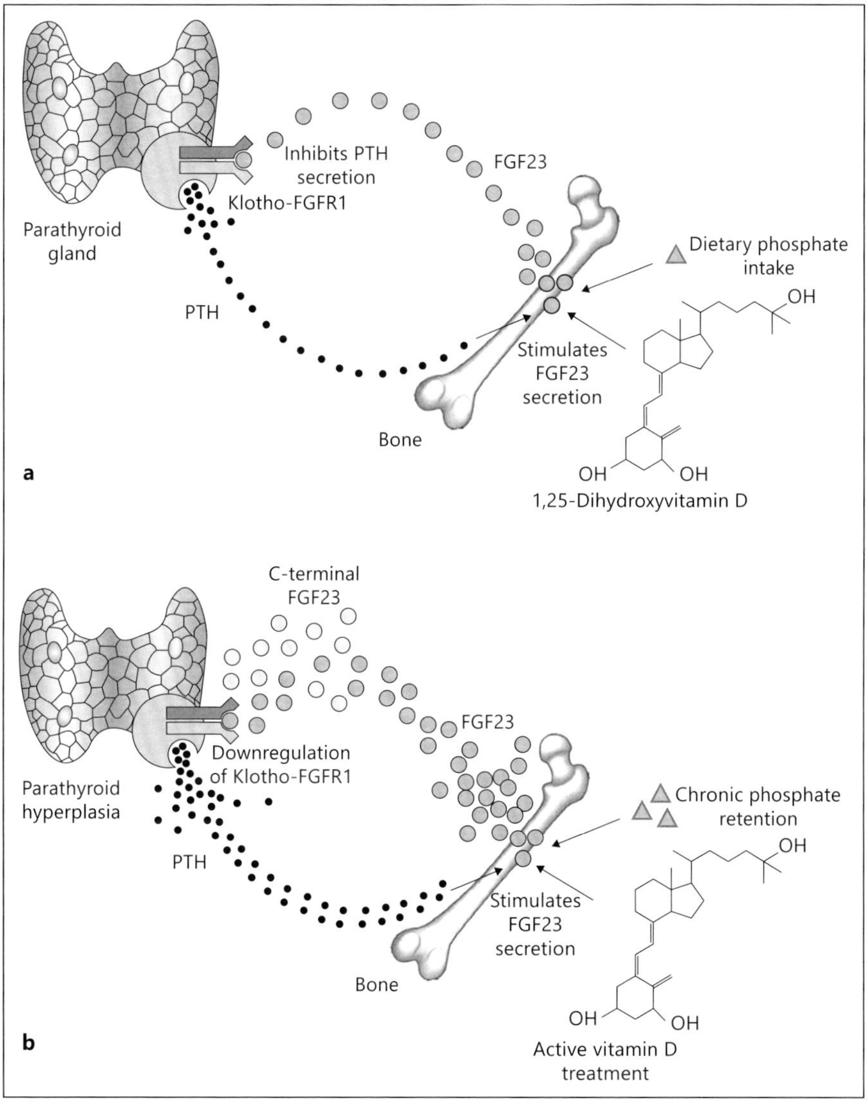

Fig. 1. Interactions between PTH and FGF23 in health and CKD. **a** In normal conditions, FGF23 secretion in the bone is stimulated in response to dietary phosphate intake, $1,25(OH)_2D$, and PTH. FGF23 in turn acts on the parathyroid to inhibit PTH secretion. **b** In patients with CKD, FGF23 secretion is markedly increased in response to chronic phosphate retention, treatment with active vitamin D, and increased PTH. The effect of FGF23 to inhibit PTH secretion is blunted, due to decreased expression of the Klotho-FGFR 1 complex in the parathyroid and increased concentrations of C-terminal FGF23 that compete with full-length FGF23 for binding to the receptor complex.

FGF23-Klotho Axis in Kidney Transplant Recipients

Hypophosphatemia due to inappropriate urinary phosphate wasting are frequently observed following successful kidney transplantation. This complication has long been attributed to persistently increased levels of PTH. Such persistent SHPT can occur even after successful kidney transplantation, presumably due to persistently decreased expression of CaSR and VDR in the nodular form of parathyroid hyperplasia [37]. However, this view cannot explain the concurrent $1,25(OH)_2D$ deficiency or the development of urinary phosphate wasting in the absence of elevated PTH levels. In this context, recent data suggest that FGF23 is the primary pathogenic factor that accounts for the syndrome of posttransplant hypophosphatemia and concomitant vitamin D deficiency [38]. It is now considered that in the pretransplant period, FGF23 secretion is extremely upregulated in order to counteract chronic phosphate retention, but fails to exert its hormonal effects in the absence of functioning kidneys. After kidney transplantation, this excess FGF23 acts on the allograft to promote phosphaturia and suppress $1,25(OH)_2D$ production.

Effects of FGF23 Independent of Klotho

Recent clinical studies of CKD patients have shown associations between high levels of FGF23 and greater risks of morbidity and mortality [39]. Although these data suggest that measurement of FGF23 may serve as a novel biomarker to guide the treatment of disordered phosphate metabolism, whether FGF23 directly contributes to poor clinical outcomes or is simply a biomarker of toxicity of other factors has been unknown. Recently, Faul et al. [40] have demonstrated a pathogenic role of FGF23 in the development of left ventricular hypertrophy through a Klotho-independent pathway. These data not only explain the high rates of left ventricular hypertrophy and mortality in patients with CKD, but also open a new avenue for research on the off-target effects of FGF23. Future studies should investigate the effects of FGF23 in other nontarget organs and examine whether treatment to reduce FGF23 provides survival benefit in patients with CKD.

Role of Soluble Klotho in SHPT

Klotho is also cleaved and released into the circulation and functions as a humoral factor. Recent experimental studies suggest that this soluble form of Klotho plays a role in the regulation of mineral homeostasis. Secreted Klotho enhances calcium reabsorption in the distal nephron by activating the transient receptor potential v-5 [41], and also functions as a phosphaturic substance via its enzymatic action on renal sodium-dependent phosphate transporter 2a [42]. More recent studies have shown that soluble Klotho directly inhibits phosphate uptake by vascular smooth muscles in vitro [43]. The role of soluble Klotho in SHPT is, however, largely unknown.

Treatment for SHPT and FGF23-Klotho Axis

On the basis of accumulating data linking high levels of FGF23 to poor outcomes [39], recent studies have focused on whether these levels can be altered by treatment for SHPT. Several clinical studies also examined the potential of FGF23 as an indicator of response to treatment for SHPT.

Phosphate Binders

Since the primary stimulus for FGF23 secretion would be high dietary phosphate intake, recent studies examined the effects of phosphate binders on serum FGF23 levels in patients with CKD. The first study showed a significant reduction in serum FGF23 levels after 4-week treatment with sevelamer hydrochloride plus calcium carbonate in dialysis patients [44]. A recent study also reported a similar reduction in serum FGF23 after 16-week treatment with lanthanum carbonate on top of the conventional treatment with calcium carbonate [45]. For patients with predialysis CKD, treatment with sevelamer hydrochloride and lanthanum carbonate both resulted in a significant reduction in serum FGF23 levels [46, 47]. Another study failed to show a significant reduction in serum FGF23 after administration of lanthanum carbonate, which may be due to a short treatment period (2 weeks) [48].

Active Vitamin D Derivatives

The stimulatory effect of active vitamin D on FGF23 was first postulated in a study of hemodialysis patients [31]. This finding is subsequently confirmed by experimental studies showing that $1,25(OH)_2D$ directly stimulates FGF23 production in osteoblasts [49]. Recent studies also showed that the response of FGF23 to active vitamin D therapy did not differ between the types of active vitamin D derivatives, including calcitriol, doxercalciferol, alfacalcidol and paricalcitol [50, 51]. The clinical implications of these findings, however, remain unclear. Further research is required to determine whether increased FGF23 by active vitamin D treatment attenuates the putative beneficial effects of active vitamin D on survival in patients with CKD [52].

Clinical studies also showed that pretreatment serum FGF23 levels predict the effectiveness of active vitamin D therapy [53] and the future development of refractory hyperparathyroidism in dialysis patients with SHPT [54]. The mechanism of this finding is not clear, but one possibility is that high levels of FGF23 at baseline was a consequence of prolonged active vitamin D administration for severe SHPT [31], which may lead to future resistance to this therapy. Another is that chronic phosphate retention, as reflected by elevated FGF23 levels, contributed to further progression of parathyroid hyperplasia. In addition, it has recently been shown that high FGF23 levels may be a marker for functional parathyroid hyperplasia as evaluated by dynamic calcium-mediated suppression of PTH levels in hemodialysis [55]. In line with this observation, we recently found a positive correlation of serum FGF23 with parathyroid gland volume, as evaluated by ultrasonography, independent of PTH levels in patients with SHPT [unpubl. data].

Calcimimetics

It has recently been shown that a treatment strategy of cinacalcet hydrochloride plus low-dose active vitamin D results in a greater decrease in FGF23 levels compared with treatment with titrated doses of active vitamin D [56]. However, in this study participants in the cinacalcet arm were withdrawn from their active vitamin D and thereafter treated with fixed low-dose active vitamin D. Because active vitamin D increases FGF23 increases serum FGF23 levels in hemodialysis patients [31], the reduction in serum FGF23 in the cinacalcet group may in part be attributed to the lower exposure to active vitamin D.

Recently, we measured serum FGF23 levels in 55 hemodialysis patients participating the clinical trial of cinacalcet hydrochloride for treating SHPT [57]. In the study, alteration of vitamin D dosage was not permitted except for the case in which serum calcium could not be managed by calcium carbonate adjustment alone. The results showed that treatment with cinacalcet, with a constant dosage of active vitamin D, resulted in significant and sustained reductions in serum FGF23. These results support the possibility that cinacalcet treatment per se leads to a reduction in FGF23 levels. Although the mechanisms responsible for the reduction in FGF23 levels during treatment with cinacalcet are uncertain, the recent finding of the stimulatory effect of PTH on FGF23 secretion [32] suggests that cinacalcet-induced reduction in serum PTH or its subsequent effects on bone metabolism contributes to the reduction in serum FGF23. Another possibility is that alterations in serum calcium or phosphate mediated the effect of cinacalcet on FGF23 reduction.

A more recent study showed that treatment with cinacalcet resulted in only small and transient reductions in serum levels of soluble Klotho, despite significant alterations in mineral and bone metabolism during this treatment [58]. The mechanisms of these effects are unknown, but the absence of associations between changes in soluble Klotho and changes in other parameters of SHPT suggests that cinacalcet has a direct effect on circulating soluble

Klotho. Whether CaSR regulates secretion of soluble Klotho from the kidney or parathyroid should be examined in future studies.

Of note, the above-mentioned studies have also shown that pretreatment with FGF23 did not predict treatment response to cinacalcet hydrochloride for SHPT [56, 57]. These findings are in contrast to the results of previous studies of active vitamin D therapy [53, 54]. It is noteworthy that such discrepancies in the predictability of FGF23 between the two treatment agents well correspond to the fact that the presence of enlarged parathyroid gland is associated with resistance to active vitamin D therapy [20] but this is not the case for cinacalcet therapy [59]. These data collectively suggest that cinacalcet is effective even in patients with severe SHPT that is or will be resistant to active vitamin D therapy.

Conclusion

Over the past decades, identification of the FGF23-Klotho axis [60–64] and the clarification of its role as a key regulator of phosphate and $1,25(OH)_2D$ metabolism has greatly reshaped our understanding of disordered mineral metabolism in CKD. In addition, there have been considerable advances in the understanding of the pathogenesis of SHPT. Recent clinical studies also reported that high levels of FGF23 are associated with morbidity and mortality as well as treatment resistance to active vitamin D, suggesting the potential of FGF23 as a novel biomarker to guide treatment of disordered phosphate metabolism in CKD. Furthermore, the recent findings of a Klotho-independent, pathogenic role of FGF23 raises the need for further research to examine whether treatment to reduce FGF23 provides survival benefit in patients with CKD. It is hoped that further research in the field of SHPT and the FGF23-Klotho axis will provide novel therapeutic approaches to improve outcomes of patients with CKD.

References

1 Potts JT: Parathyroid hormone: past and present. J Endocrinol 2005;187:311–325.
2 Silver J, Naveh-Many T: Phosphate and the parathyroid. Kidney Int 2009;75:898–905.
3 Shimada T, Mizutani S, Muto T, Yoneya T, Hino R, Takeda S, Takeuchi Y, Fujita T, Fukumoto S, Yamashita T: Cloning and characterization of FGF23 as a causative factor of tumor-induced osteomalacia. Proc Natl Acad Sci USA 2001;98:6500–6505.
4 Ben-Dov IZ, Galitzer H, Lavi-Moshayoff V, Goetz R, Kuro-o M, Mohammadi M, Sirkis R, Naveh-Many T, Silver J: The parathyroid is a target organ for FGF23 in rats. J Clin Invest 2007;117:4003–4008.
5 Drueke TB: Cell biology of parathyroid gland hyperplasia in chronic renal failure. J Am Soc Nephrol 2000;11:1141–1152.

6 Fukagawa M, Nakanishi S, Kazama JJ: Basic and clinical aspects of parathyroid hyperplasia in chronic kidney disease. Kidney Int 2006;70(suppl 102): S3–S7.
7 Palmer SC, Hayen A, Macaskill P, Pellegrini F, Craig JC, Elder GJ, Strippoli GF: Serum levels of phosphorus, parathyroid hormone, and calcium and risks of death and cardiovascular disease in individuals with chronic kidney disease: a systematic review and meta-analysis. JAMA 2011;305:1119–1127.
8 Komaba H, Fukagawa M: FGF23-parathyroid interaction: implications in chronic kidney disease. Kidney Int 2010;77:292–298.
9 Martin KJ, González EA: Metabolic bone disease in chronic kidney disease. J Am Soc Nephrol 2007;18:875–885.
10 Bricker NS: On the pathogenesis of the uremic state. An exposition of the 'trade-off hypothesis'. N Engl J Med 1972;286:1093–1099.
11 Slatopolsky E, Caglar S, Pennell JP, Taggart DD, Canterbury JM, Reiss E, Bricker NS: On the pathogenesis of hyperparathyroidism in chronic experimental renal insufficiency in the dog. J Clin Invest 1971;50:492–499.
12 Portale AA, Booth BE, Halloran BP, Morris RCJ: Effect of dietary phosphorus on circulating concentrations of 1,25-dihydroxyvitamin D and immunoreactive parathyroid hormone in children with moderate renal insufficiency. J Clin Invest 1984;73:1580–1589.
13 Slatopolsky E, Finch J, Denda M, Ritter C, Zhong M, Dusso A, MacDonald PN, Brown AJ: Phosphorus restriction prevents parathyroid gland growth. High phosphorus directly stimulates PTH secretion in vitro. J Clin Invest 1996;97:2534–2540.
14 Kilav R, Silver J, Naveh-Many T: A conserved cis-acting element in the parathyroid hormone 3′-untranslated region is sufficient for regulation of RNA stability by calcium and phosphate. J Biol Chem 2001;276:8727–8733.
15 Nykjaer A, Dragun D, Walther D, Vorum H, Jacobsen C, Herz J, Melsen F, Christensen EI, Willnow TE: An endocytic pathway essential for renal uptake and activation of the steroid 25-(OH) vitamin D_3. Cell 1999;96:507–515.
16 Wang Q, Palnitkar S, Parfitt AM: The basal rate of cell proliferation in normal human parathyroid tissue: implications for the pathogenesis of hyperparathyroidism. Clin Endocrinol (Oxf) 1997;46:343–349.
17 Tominaga Y, Tanaka Y, Sato K, Nagasaka T, Takagi H: Histopathology, pathophysiology, and indications for surgical treatment of renal hyperparathyroidism. Semin Surg Oncol 1997;13:78–86.
18 Kifor O, Moore FD Jr, Wang P, Goldstein M, Vassilev P, Kifor I, Hebert SC, Brown EM: Reduced immunostaining for the extracellular Ca^{2+}-sensing receptor in primary and uremic secondary hyperparathyroidism. J Clin Endocrinol Metab 1996;81:1598–1606.
19 Fukuda N, Tanaka H, Tominaga Y, Fukagawa M, Kurokawa K, Seino Y: Decreased 1,25-dihydroxyvitamin D_3 receptor density is associated with a more severe form of parathyroid hyperplasia in chronic uremic patients. J Clin Invest 1993;92:1436–1443.
20 Fukagawa M, Kitaoka M, Yi H, Fukuda N, Matsumoto T, Ogata E, Kurokawa K: Serial evaluation of parathyroid size by ultrasonography is another useful marker for the long-term prognosis of calcitriol pulse therapy in chronic dialysis patients. Nephron 1994;68: 221–228.
21 Arcidiacono MV, Sato T, Alvarez-Hernandez D, Yang J, Tokumoto M, Gonzalez-Suarez I, Lu Y, Tominaga Y, Cannata-Andia J, Slatopolsky E, Dusso AS: EGFR activation increases parathyroid hyperplasia and 1,25(OH)2D resistance in kidney disease. J Am Soc Nephrol 2008;19:310–320.
22 Fukagawa M, Kazama JJ, Shigematsu T: Skeletal resistance to PTH as a basic abnormality underlying uremic bone diseases. Am J Kidney Dis 2001;38:S152–S155.
23 Nii-Kono T, Iwasaki Y, Uchida M, Fujieda A, Hosokawa A, Motojima M, Yamato H, Kurokawa K, Fukagawa M: Indoxyl sulfate induces skeletal resistance to parathyroid hormone in cultured osteoblastic cells. Kidney Int 2007; 71:738–743.
24 Iwasaki Y, Yamato H, Nii-Kono T, Fujieda A, Uchida M, Hosokawa A, Motojima M, Fukagawa M: Administration of oral charcoal adsorbent (AST-120) suppresses low-turnover bone progression in uraemic rats. Nephrol Dial Transplant 2006;21:2768–2774.
25 Kuro-o M, Matsumura Y, Aizawa H, Kawaguchi H, Suga T, Utsugi T, Ohyama Y, Kurabayashi M, Kamane T, Kume E, Iwasaki H, Iida A, Shiraki-Iida T, Nishikawa S, Nagai R, Nabeshima Y: Mutation of the mouse klotho gene leads to a syndrome resembling ageing. Nature 1997;390:45–51.

26 Kurosu H, Ogawa Y, Miyoshi M, Yamamoto M, Nandi A, Rosenblatt KP, Baum MG, Schiavi S, Hu MC, Moe OW, Kuro-o M: Regulation of fibroblast growth factor-23 signaling by Klotho. J Biol Chem 2006;281:6120–6123.
27 Urakawa I, Yamazaki Y, Shimada T, Iijima K, Hasegawa H, Okawa K, Fujita T, Fukumoto S, Yamashita T: Klotho converts canonical FGF receptor into a specific receptor for FGF23. Nature 2006;444:770–774.
28 Shigematsu T, Kazama JJ, Yamashita T, Fukumoto S, Hosoya T, Gejyo F, Fukagawa M: Possible involvement of circulating fibroblast growth factor 23 in the development of secondary hyperparathyroidism associated with renal insufficiency. Am J Kidney Dis 2004;44: 250–256.
29 Gutierrez O, Isakova T, Rhee Shah A, Holmes J, Collerone G, Juppner H, Wolf M: Fibroblast growth factor-23 mitigates hyperphosphatemia but accentuates calcitriol deficiency in chronic kidney disease. J Am Soc Nephrol 2005;16:2205–2215.
30 Hasegawa H, Nagano N, Urakawa I, Yamazaki Y, Iijima K, Fujita T, Yamashita T, Fukumoto S, Shimada T: Direct evidence for a causative role of FGF23 in the abnormal renal phosphate handling and vitamin D metabolism in rats with early-stage chronic kidney disease. Kidney Int 2010;78:975–980.
31 Nishi H, Nii-Kono T, Nakanishi S, Yamazaki Y, Yamashita T, Fukumoto S, Ikeda K, Fujimori A, Fukagawa M: Intravenous calcitriol therapy increases serum concentration of fibroblast growth factor 23 dialysis patients with secondary hyperparathyroidism. Nephron Clin Prac 2005;101:c94–c99.
32 Lavi-Moshayoff V, Wasserman G, Meir T, Silver J, Naveh-Many T: PTH increases FGF23 gene expression and mediates the high-FGF23 levels of experimental kidney failure: a bone parathyroid feedback loop. Am J Physiol Renal Physiol 2010;299:F882–F889.
33 Sato T, Tominaga Y, Ueki T, Goto N, Matsuoka S, Katayama A, Haba T, Uchida K, Nakanishi S, Kazama JJ, Gejyo F, Yamashita T, Fukagawa M: Total parathyroidectomy reduces elevated circulating fibroblast growth factor 23 in advanced secondary hyperparathyroidism. Am J Kidney Dis 2004; 44:481–487.
34 Komaba H, Goto S, Fujii H, Hamada Y, Kobayashi A, Shibuya K, Tominaga Y, Otsuki N, Nibu K, Nakagawa K, Tsugawa N, Okano T, Kitazawa R, Fukagawa M: Depressed expression of Klotho and FGF receptor 1 in hyperplastic parathyroid glands from uremic patients. Kidney Int 2010;77:232–238.
35 Galitzer H, Ben-Dov IZ, Silver J, Naveh-Many T: Parathyroid cell resistance to fibroblast growth factor 23 in secondary hyperparathyroidism of chronic kidney disease. Kidney Int 2010;77:211–218.
36 Goetz R, Nakada Y, Hu MC, Kurosu H, Wang L, Nakatani T, Shi M, Eliseenkova AV, Razzaque MS, Moe OW, Kuro-o M, Mohammadi M: Isolated C-terminal tail of FGF23 alleviates hypophosphatemia by inhibiting FGF23-FGFR-Klotho complex formation. Proc Natl Acad Sci USA 2010;107:407–412.
37 Taniguchi M, Tokumoto M, Matsuo D, Motoyama K, Sugitani A, Kuroki S, Yotsueda H, Tsuruya K, Hirakata H, Iida M: Persistent hyperparathyroidism in renal allograft recipients: vitamin D receptor, calcium-sensing receptor, and apoptosis. Kidney Int 2006;70: 363–370.
38 Bhan I, Shah A, Holmes J, Isakova T, Gutierrez O, Burnett SM, Jüppner H, Wolf M: Post-transplant hypophosphatemia: tertiary 'hyper-phosphatoninism'? Kidney Int 2006;70: 1486–1494.
39 Gutierrez OM, Mannstadt M, Isakova T, Rauh-Hain JA, Tamez H, Shah A, Smith K, Lee H, Thadhani R, Jüppner H, Wolf M: Fibroblast growth factor 23 and mortality among patients undergoing hemodialysis. N Engl J Med 2008;359:584–592.
40 Faul C, Amaral AP, Oskouei B, Hu MC, Sloan A, Isakova T, Gutierrez OM, Aguillon-Prada R, Lincoln J, Hare JM, Mundel P, Morales A, Scialla J, Fischer M, Soliman EZ, Chen J, Go AS, Rosas SE, Nessel L, Townsend RR, Feldman HI, John Sutton MS, Ojo A, Gadegbeku C, Di Marco GS, Reuter S, Kentrup D, Tiemann K, Brand M, Hill JA, Moe OW, Kuro-o M, Kusek JW, Keane MG, Wolf M: FGF23 induces left ventricular hypertrophy. J Clin Invest 2011;121:4393–4408.
41 Chang Q, Hoefs S, van der Kemp AW, Topala CN, Bindels RJ, Hoenderop JG: The beta-glucuronidase klotho hydrolyzes and activates the TRPV5 channel. Science 2005;310: 490–493.

Razzaque MS (ed): Phosphate and Vitamin D in Chronic Kidney Disease.
Contrib Nephrol. Basel, Karger, 2013, vol 180, pp 124–137 (DOI: 10.1159/000346792)

FGF23-Induced Hypophosphatemia Persists in *Hyp* Mice Deficient in the WNT Coreceptor Lrp6

Kazuyoshi Uchihashi[a] · Teruyo Nakatani[a] · Regina Goetz[b] · Moosa Mohammadi[b] · Xi He[c] · Mohammed S. Razzaque[a]

[a]Department of Oral Medicine, Infection and Immunity, Harvard School of Dental Medicine and [c]Department of Neurobiology, Harvard Medical School, Boston, Mass., and [b]Department of Pharmacology, New York University School of Medicine, New York, N.Y., USA

Abstract

Deregulated phosphate homeostasis can lead to a wide range of disorders, including myopathy, cardiac dysfunction, and skeletal abnormalities. Therefore, characterization of the molecular regulation of phosphate metabolism is of pathophysiological and clinical significance. *Hyp* mouse is the model for human X-linked hypophosphatemia which is due to mutations that inactivate the endopeptidases of the X chromosome (PHEX). PHEX inactivation leads to increased serum levels of fibroblast growth factor 23 (FGF23), a phosphaturic hormone that induces excessive renal phosphate excretion and severe hypophosphatemia. The expression of WNT signaling components is increased in *Hyp* mice. To determine the potential role of WNT signaling in FGF23-mediated hypophosphatemia, we cross-bred *Hyp* mice with mice deficient in the WNT coreceptor low-density lipoprotein receptor-related protein 6 (*Lrp6*) to generate *Hyp* and *Lrp6* double mutant mice (*Hyp/Lrp6*). Like *Hyp* mice, *Hyp/Lrp6* double mutants maintained high serum levels of FGF23, and accordingly exhibited hypophosphatemia to the same degree as the *Hyp* mice did, indicating that genetically reducing WNT signaling does not impact FGF23-induced phosphaturia. Moreover, similar to *Hyp* mice, the *Hyp/Lrp6* double mutants also exhibited reduced mineralization of the bone, further supporting that reduced WNT signaling does not affect the chronic phosphate wasting caused by excess FGF23 in these mice. In further support of our finding, injection of bioactive FGF23 protein into *Lrp6* mutant mice reduced serum phosphate levels to a similar degree as FGF23 injection into wild-type mice. Our in vivo studies provide genetic and pharmacological evidence for a WNT-independent function of FGF23 in the regulation of phosphate homeostasis.

Determining the molecular regulation of phosphate metabolism is of great pathophysiological and clinical significance, as altered phosphate balance can induce a wide range of disorders, including myopathy, cardiac and skeletal dysfunctions [1–6]. Fibroblast growth factor 23 (FGF23) is a phosphaturic hormone that plays a key role in the regulation of phosphate metabolism [7–12]. This is reflected by the profound clinical manifestations of either abnormally high or abnormally low bioactivity of FGF23. Increased FGF23 activity causes renal phosphate wasting and leads to the development of hypophosphatemic rickets and osteomalacia, whereas reduced FGF23 activity results in enhanced renal phosphate retention and manifests in hyperphosphatemic rickets and ectopic calcifications, among others [6–9]. Recent studies have pointed to a possible role of WNT signaling in FGF23-mediated functions, and have also suggested that WNT signaling can stimulate the promoter activity of the FGF23 gene in osteoblasts [13, 14].

The canonical pathway of WNT signaling has been implicated in numerous essential biological functions, ranging from embryogenesis to organogenesis [15–17]. In mammals, canonical WNT signaling is initiated through the binding of WNT to Frizzled receptors and the low-density lipoprotein receptor-related protein-5 or protein-6 (Lrp5/6), to stabilize beta-catenin, which induces downstream target genes via the T cell factor and lymphoid enhancer binding factor [14–18]. According to the cellular requirement, the secreted molecule Dickkopf 1 (Dkk-1) can fine-tune the WNT signaling activity by a negative feedback mechanism [18]; Dkk-1 antagonizes WNT signaling by disrupting the molecular interaction between WNT and Lrp5/6 [19–21]. Studies using genetic or pharmacological approaches have convincingly shown that inactivating the function of either of the two WNT coreceptors, Lrp5 or Lrp6, can significantly reduce WNT activities to influence bone growth and maintenance [22, 23]. Disruption of *Lrp6* activity in mice resulted in developmental anomalies [24], including limb patterning defects. Importantly, some of the *Lrp6* mutant phenotypes are similar to the changes observed in embryos lacking *Wnt1*, *Wnt3a* and *Wnt7a* activity [25–27]. Since homozygous *Lrp6–/–* mice are embryonic lethal and display numerous phenotypes including limb defects and axial truncation [24], we used heterozygous *Lrp6+/–* mice as a model system to reduce WNT activities from *Hyp* mice.

PHEX is a phosphate-regulating gene with homology to endopeptidases of the X chromosome, and is a membrane-bound ectoenzyme expressed in bone cells. Mutations in the *PHEX* gene are associated with X-linked hypophosphatemia, a renal phosphate wasting disorder in humans [28–30]. *Hyp* mice are the murine model of X-linked hypophosphatemia, with an inactivating mutation in the *PHEX* gene. Similarly as in the human disease, increased activity of FGF23 in the *Hyp* mice leads to hypophosphatemia, due to excessive urinary loss of phosphate [29, 31–33]; such renal loss of phosphate eventually leads to rickets and/or osteomalacia. In this study, we generated and characterized *Hyp/Lrp6* double mutant mice to determine the in vivo role of WNT signaling in FGF23-mediated systemic regulation of phosphate homeostasis.

Materials and Methods

Generation of Hyp/Lrp6 Double Mutant Mice
We crossbred *Hyp* mutants (Jackson Laboratory, Me., USA) with heterozygous *Lrp6* mutants to obtain *Hyp/Lrp6* double mutants. Routine PCR using genomic DNA extracted from tail clips was performed for genotyping the various groups of mice [31, 34, 35]. Since homozygous *Lrp6–/–* are embryonic lethal and display numerous phenotypes including limb defects and axial truncation [24], we used heterozygous *Lrp6+/–* mice for our study. The *Lrp6+/–* mice have a reduction in bone density and are viable, fertile and usually do not display any obvious phenotype, except occasional kinked/curly tail, spina bifida and imperforated vagina. The following primer sequences were used to genotype *Lrp6* mutant mice (LRP6 forward primer: CAG GCA TGT AGC CCT TGG AG; LRP6 reverse primer: ACT ACA AGC CCT GCA CTG CC and LRP6 Neo reverse primer: GTA GAG TTC CCA GGA GGA GCC). The PCR product for wild-type Lrp6 mice was detected as 285 bp, for heterozygous Lrp6 as 385 and 285 bp, and for homozygous Lrp6 as 385 bp. Mice were maintained in accordance with the NIH Guide for the Care and Use of Laboratory Animals and were employed using protocols approved by the institution's subcommittee on animal care (IACUC). The total body weight of wild-type, *Hyp*, *Lrp6+/–* and *Hyp/Lrp6* double mutant mice were taken weekly starting at 3 weeks of age until 20 weeks. At least five or more mice from each genotype were sacrificed at around 8 weeks and long bones, including femur and tibia were isolated for radiological imaging. Plain radiographs were taken (29 kV for 10 s) using an X-ray apparatus (Micro 50; Micro Focus Imaging). Films were scanned (scanjet 5530; Hewlett-Packard) and saved.

Serum Measurements
Blood was obtained from wild-type, *Hyp*, *Lrp6+/--* and *Hyp/Lrp6* double mutant mice by cheek pouch bleeding at different time points. Serum was isolated by centrifugation at 3,000 *g* for 10 min and was stored at –80 °C until use. Serum phosphorus and calcium were determined by colorimetric measurements using the Stanbio Phosphorus Liqui-UV Test and the Calcium (Arsenazo) LiquiColor Test, respectively. The serum level of FGF23 was measured with ELISA using a commercial kit (Kainos Laboratories, Tokyo, Japan), as described in a previous publication [31].

Quantitative Real-Time PCR
Total RNA was isolated from the kidneys of wild-type, *Hyp*, *Lrp6+/–* and *Hyp/Lrp6* double mutant mice, and used to measure relative expression of *NaPi-2a* mRNA by real-time PCR, as described earlier. Real-time PCR was performed in duplicate, and all reactions

were controlled by standards, as detailed in earlier studies [36, 37]. The relative quantity of mRNA was calculated by normalizing the CT (threshold cycle value) of *NaPi-2a* to the CT of the housekeeping gene GAPDH.

Treatment of Mice with Recombinant Bioactive FGF23

Wild-type and *Lrp6+/–* mutant mice were injected intraperitoneally with vehicle, or recombinant bioactive FGF23 protein (10 μg per mouse) [38]. Blood was collected by cheek-pouch bleeding prior to the injections and at 9 h postinjection. Isolated serum was used to determine phosphate levels in an animal pre- and postinjection, using the Stanbio Liqui-UV kit [39].

Statistics

Statistically significant differences between groups were evaluated by either Student's t test or Mann-Whitney U test for a comparison between the groups. All values are expressed as mean ± SE. $p < 0.05$ was considered to be statistically significant. All analyses were performed using StatView.

Results

Generation of Hyp/Lrp6 Double Mutant Mice

To elucidate the consequences of reducing WNT activity from *Hyp* mice in vivo, we generated a new *Hyp* mouse model deficient in *Lrp6* function (*Hyp/Lrp6* double mutant mice) (fig. 1a). The generated double mutant mice were viable throughout their adulthood but were smaller than wild-type mice (fig. 1a). At birth, *Hyp/Lrp6* double mutant mice were indistinguishable from littermates of other genotypes. However, at 8 weeks of age, double mutant mice (24.1 ± 0.3 g) were smaller significantly ($p < 0.01$) than wild-type mice (29.2 ± 0.8 g) and *Lrp6+/–* mice (28.5 ± 2.2 g), but similar in weight as *Hyp* mice (23.5 ± 0.5 g) (fig. 1a, b). At 16 weeks of age, *Hyp/Lrp6* double mutant mice (28.9 ± 0.4 g), like *Hyp* mice (26.6 ± 0.5 g), still lagged behind their wild-type (38.9 ± 2.0 g) and *Lrp6+/–* (35.1 ± 3.8 g) littermates with regard to body weight (fig. 1b).

Serum Phosphate and Calcium Levels of Hyp/Lrp6 Double Mutant Mice

We measured serum phosphate and calcium levels in the various genotypes. As expected, *Hyp* mice showed reduced serum phosphate levels (4.3 ± 0.2 mg/dl) compared to wild-type mice (6.8 ± 0.3 mg/dl) at around 6–10 weeks of age. Serum phosphate of *Hyp/Lrp6* double mutant mice (3.9 ± 0.4 mg/dl) was similarly low as that of *Hyp* mice at this age, whereas serum phosphate concentrations of age-matched *Lrp6+/–* mice (6.0 ± 0.2 mg/dl) were within the normal range (fig. 2a, left panel). A similar pattern of serum phosphate levels was also noted in mice of 11 to 16 weeks of age. Compared to wild-type mice (7.0 ± 0.4 mg/dl), *Hyp/Lrp6* double mutant mice were hypophosphatemic at 11 to 16 weeks of age,

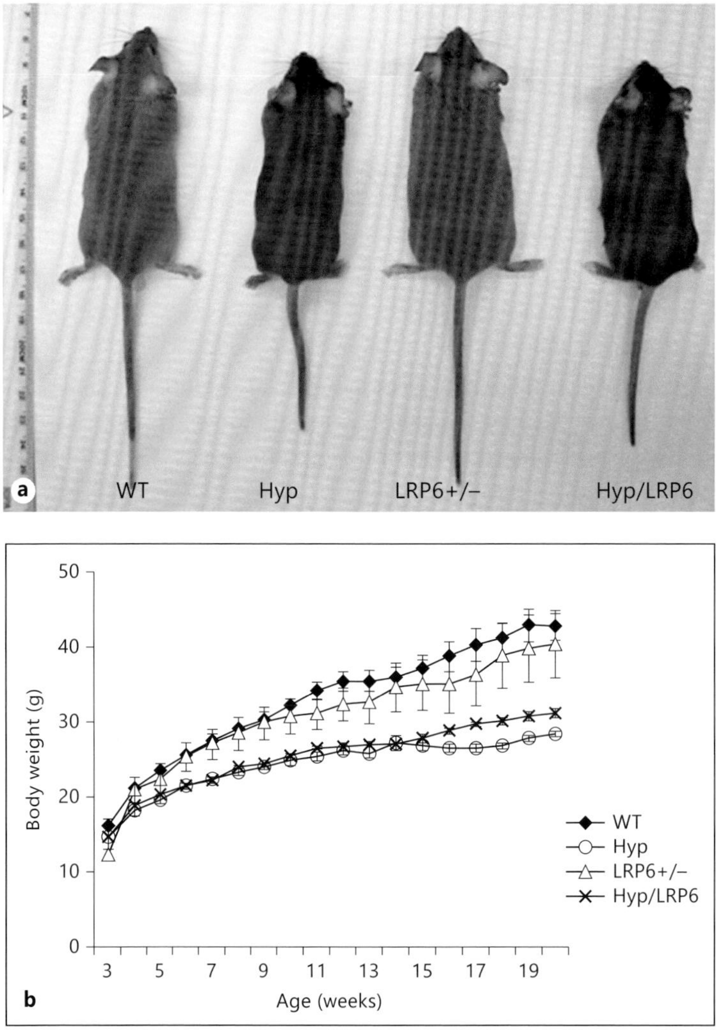

Fig. 1. Gross phenotype of *Hyp/Lrp6* double mutant mice. **a** Gross appearance of male wild-type (WT), *Hyp*, *Lrp6+/–*, and *Hyp/Lrp6* double mutant mice at 8 weeks of age. **b** Growth curve of WT (n = 6), *Hyp* (n = 8), *Lrp6+/–* (n = 7) and *Hyp/Lrp6* double mutant mice (n = 5). Note that the physical appearance of *Hyp/Lrp6* double mutant mice is identical with that of *Hyp* mice; in particular, mice of these two genotypes are smaller in size than WT or *Lrp6+/–* mice.

as were *Hyp* mice. By contrast, *Lrp6+/–* mice remained normophosphatemic (6.6 ± 0.3 mg/dl). Importantly, the low serum phosphate levels in *Hyp/Lrp6* double mutant mice (4.4 ± 0.1 mg/dl) were similar to those seen in *Hyp* mice (5.0 ± 0.5 mg/dl) at 11–16 weeks of age (fig. 2a, right panel). As in *Hyp* mice, renal expression of NaPi-2a was reduced in *Hyp/Lrp6* double mutant mice compared

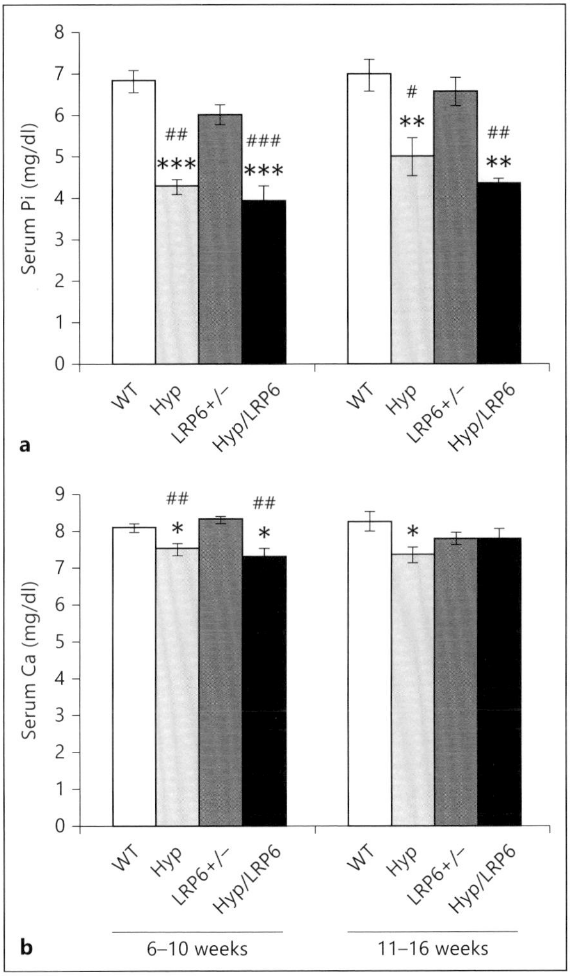

Fig. 2. Serum phosphate and serum calcium in *Hyp/Lrp6* double mutant mice. **a** Serum phosphate levels of wild-type (WT), *Hyp*, *Lrp6+/–*, and *Hyp/Lrp6* double mutant mice at 6–10 weeks and 11–16 weeks of age. Note that the serum phosphate levels of *Hyp/Lrp6* double mutant mice are significantly below the normal level, and are as low as those of *Hyp* mice. For example, at 6–10 weeks of age, serum phosphate of *Hyp/Lrp6* double mutant mice is 3.9 ± 0.4 (n = 7) compared to 4.3 ± 0.2 in *Hyp* mice (n = 9) and 6.8 ± 0.3 in WT mice (n = 7). **b** Serum calcium levels of WT, *Hyp*, *Lrp6+/–*, and *Hyp/Lrp6* double mutant mice at 6–10 weeks and 11–16 weeks of age. Note that the serum calcium of *Hyp/Lrp6* double mutant mice (7.3 ± 0.3, n = 7) is slightly lower than normal (8.1 ± 0.14, n = 7) at 6–10 weeks of age, as also seen in *Hyp* mice (7.5 ± 0.2, n = 9). * $p < 0.05$, ** $p < 0.01$ and *** $p < 0.001$ vs. WT; # $p < 0.05$, ## $p < 0.01$ and ### $p < 0.001$ vs. *Lrp6+/–*.

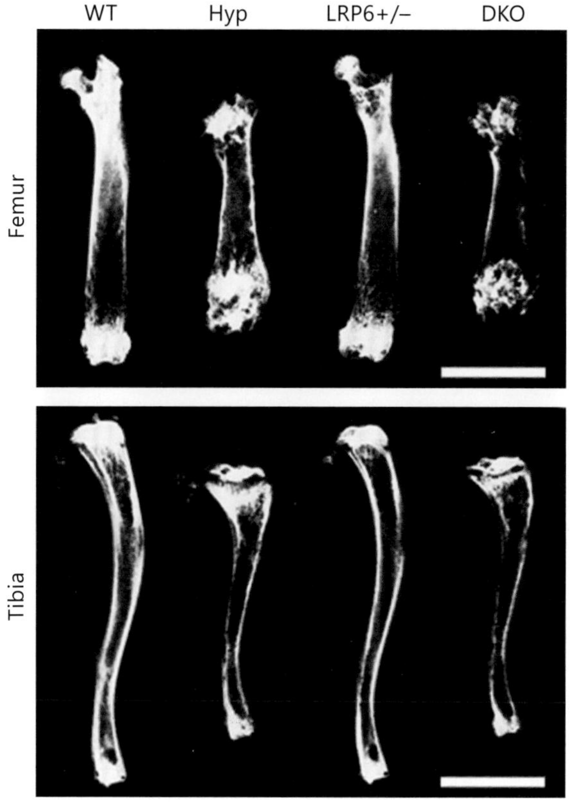

Fig. 3. X-ray analysis of bone tissue in *Hyp/Lrp6* double mutant mice. Shown are X-ray images of femur and tibia from male wild-type (WT), *Hyp*, *Lrp6+/–*, and *Hyp/Lrp6* double mutant mice at 8 weeks of age. Like *Hyp* mice, the *Hyp/Lrp6* double mutant mice show reduced skeletal mineralization of bone with the ricketic widening of the growth plate. Please note that the skeletal anomalies are more pronounced in *Hyp/Lrp6* double mutant mice than in *Hyp* mice. Scale bar = 5 mm.

to wild-type littermates, which likely accounted for the hypophosphatemia (fig. 5). There was no change in NaPi-2a expression in the kidneys of *Lrp6+/–* mice compared to wild-type mice (fig. 5).

Concomitant with the hypophosphatemia, *Hyp/Lrp6* double mutant mice (7.3 ± 0.3 mg/dl), like *Hyp* mice (7.5 ± 0.2 mg/dl), had slightly lower serum calcium levels than wild-type mice (8.1 ± 0.1 mg/dl) at around 6–10 weeks of age (fig. 2b, left panel). By contrast, the serum calcium level of age-matched *Lrp6+/–* mice was within the normal range (8.3 ± 0.1 mg/dl; fig. 2b, left panel), as expected based on their normophosphatemic status. Reduced serum calcium and phosphate levels both in *Hyp/Lrp6* double mutant mice and in *Hyp* mice were associated with defective skeletal mineralization compared to wild-type litter-

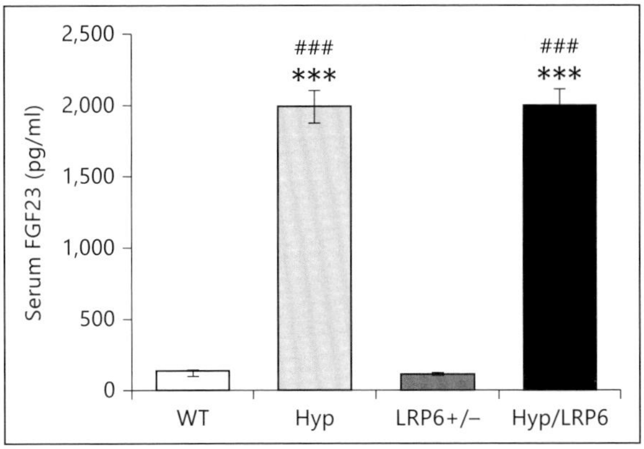

Fig. 4. Serum FGF23 in *Hyp/Lrp6* double mutant mice. Average levels of serum FGF23 in wild-type (WT), *Hyp, Lrp6+/–* and *Hyp/Lrp6* double mutant mice at 8 weeks of age are shown. Note that serum FGF23 of *Hyp/Lrp6* double mutant mice (1,997 ± 139 pg/ml, n = 5) is elevated by nearly 20-fold above normal (137 ± 9 pg/ml, n = 5). The increase in serum FGF23 is similar in magnitude to that seen in *Hyp* mice (1,989 ± 139 pg/ml, n = 5). *** p < 0.001 vs. WT, ### p < 0.001 vs. *Lrp6+/–*.

mates (fig. 3). In *Lrp6+/–* mice, bone mineralization was somewhat reduced compared to wild-type mice (fig. 3). Compared to wild-type mice, the bones of *Hyp/Lrp6* double mutant mice were short and thick, and showed the typical features of rickets such as widening of the epiphysis and cupping of the metaphysis, all of which resembling the phenotype of *Hyp* mice, except that the skeletal anomalies were more marked in *Hyp/Lrp6* double mutant mice than in *Hyp* mice (fig. 3). No such bone phenotype was seen in the mice deficient in *Lrp6* only (fig. 3).

Serum FGF23 Levels in Hyp/Lrp6 Double Mutant Mice

As expected, serum FGF23 levels were extremely high in the *Hyp* mice (1,989.4 ± 139.6 pg/ml) at 8 weeks of age compared to wild-type littermates (137.2 ± 9.1 pg/ml; fig. 4). *Hyp/Lrp6* double mutant mice showed a similar increase of serum FGF23 (1,997.2 ± 139.7 pg/ml) as *Hyp* mice at this age (fig. 4). In mice deficient in *Lrp6* only, serum levels of FGF23 were within normal range (111.9 ± 11.7 pg/ml; fig. 4). We have previously shown that increased serum levels of FGF23 are associated with the hypophosphatemia in *Hyp* mice, since serum phosphate can be increased by blocking FGF23 activity in these mice [40]. Thus, the lack of a difference between *Hyp/Lrp6* double mutant mice and *Hyp* mice in regard to serum FGF23 and concomitant hypophosphatemia suggests that the phosphaturic activity of FGF23 is a WNT-independent process.

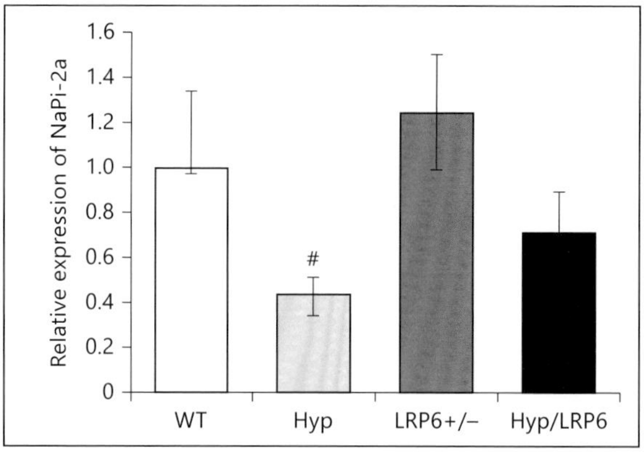

Fig. 5. Renal NaPi-2a gene expression in *Hyp/Lrp6* double mutant mice. Shown are mRNA levels of NaPi-2a in the kidneys of *Hyp, Lrp6+/–*, and *Hyp/Lrp6* double mutant mice relative to those of wild-type (WT) mice. For each tissue sample, NaPi-2a mRNA expression was normalized to that of the house-keeping gene GAPDH. Note that mRNA levels for NaPi-2a are reduced in *Hyp/Lrp6* double mutant mice similarly as in *Hyp* mice when compared to WT or Lrp6+/– mice. # $p<0.05$, *vs. Lrp6+/–*.

Effects of FGF23 Protein Injection into Lrp6-Deficient Mice

To further explore whether suppressing WNT signaling activity affects FGF23-mediated regulation of phosphate metabolism, we injected recombinant bioactive FGF23 protein into *Lrp6*-deficient mice. FGF23 administration to wild-type mice served as a control. As expected, exogenous FGF23 significantly reduced serum phosphate levels in wild-type mice (preinjection 6.7 ± 0.2 mg/dl vs. postinjection 5.0 ± 0.4mg/dl; $p < 0.05$; fig. 6). Similarly, FGF23 treatment led to a significant decrease of serum phosphate levels in *Lrp6*-deficient mice (preinjection 6.8 ± 0.3 mg/dl vs. postinjection 5.5 ± 0.3 mg/dl; $p < 0.05$; fig. 6). There was no major difference in the magnitude of the effect between *Lrp6* mice and wild-type mice, which suggests that suppressing WNT signaling activity by genetically reducing *Lrp6* activity does not affect the phosphaturic activity of FGF23.

Discussion

The expression of WNT signaling components is increased in *Hyp* mice [14], suggesting a potential role for WNT signaling in the regulation of phosphate homeostasis by FGF23 [31, 32]. We have utilized two complementary approaches to determine the potential role of WNT signaling in the phosphaturic activity of FGF23 in vivo. The first is the genetic approach of generating *Hyp/Lrp6* dou-

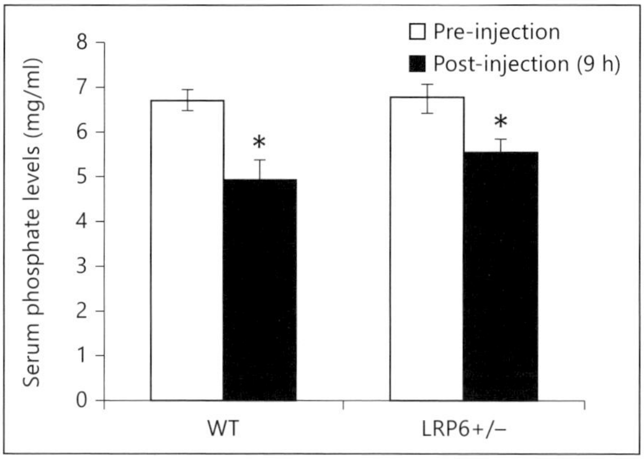

Fig. 6. Hypophosphatemic effect of exogenous FGF23 in *Lrp6*-deficient mice. Shown are serum phosphate levels before and 9 h after the last of two injections of recombinant bioactive FGF23 protein. Note that exogenous FGF23 lowers serum phosphate in *Lrp6*-deficient mice to a similar degree as in wild-type (WT) mice [*Lrp6+/–* mice (n=6): pre-injection 6.8±0.3 mg/dl vs. post-injection 5.5±0.3 mg/dl; WT mice (n=5): pre-injection 6.7±0.2 mg/dl vs. post-injection 5±0.4 mg/dl]. *p<0.05 *vs.* corresponding pre-injection value.

ble mutant mice by ablating *Lrp6* function from *Hyp* mice. The second is the pharmacological approach of treating *Lrp6+/–* mice with bioactive FGF23 protein and measuring its effects on phosphate metabolism in reduced of WNT signaling activity.

Hyp mice produce excessive amounts of FGF23 in the bone, leading to accumulation of FGF23 in serum. Such increased serum levels of FGF23 induce chronic hypophosphatemia in *Hyp* mice, owing to persistent suppression of phosphate reabsorption in the renal proximal tubule [29, 31–33]. Suppressing WNT signaling by reducing *Lrp6* function from *Hyp* mice did not change the systemic phosphate metabolism. Specifically, the hypophosphatemia was not altered by ablating *Lrp6* function from *Hyp* mice, suggesting that genetic manipulation of WNT signaling cannot affect phosphate metabolism in these mice. Further studies are needed to rule out the compensatory effects of other components of WNT signaling in *Hyp/Lrp6* double mutant mice. We then tested whether suppressing WNT signaling by ablating *Lrp6* function from otherwise normal mice affected the phosphaturic activity of exogenous FGF23. Remarkably, exogenous FGF23 exhibited a similar phosphaturic effect in *Lrp6*-deficient mice as in wild-type littermates, which clearly suggests that suppressing WNT signaling activity does not affect the regulation of systemic phosphate metabolism by FGF23. Taken together, our in vivo results suggest that the phosphaturic activity of FGF23 is a WNT-independent process.

The in vivo mechanism by which FGF23 exerts its phosphate lowering functions requires further investigation, although α-Klotho plays a vital role in such regulation. In support of such a role, the severe FGF23-induced hypophosphatemic phenotype of *Hyp* mice can be reversed to hyperphosphatemia when α-Klotho function is eliminated from *Hyp* mice. In other words, in spite of significantly increased serum levels of FGF23, without α-Klotho function, FGF23 activity is disabled [31]. Similarly, injection of bioactive FGF23 protein into either *α-klotho* knockout mice or *Fgf23/α-klotho* double knockout mice fails to lower serum phosphate levels, whereas in wild-type and *Fgf23* knockout mice, exogenous FGF23 reduces serum phosphate levels [35]. Together, these results provide compelling evidence that FGF23 does not have an α-Klotho-independent role in the systemic regulation of phosphate homeostasis. Although recent studies have established that FGF23 forms a ternary complex with cognate FGF receptor and α-Klotho to exert its phosphate-lowering effects [31, 35, 40–42], the signaling network involved in mediating the phosphaturic activity of FGF23 requires additional study. Using *Hyp* mice as an in vivo model system, our results of this study indicate that FGF23 can exert its phosphate lowering effects without involving WNT signaling. Despite the crucial importance of maintaining physiological mineral ion balance, our knowledge of the in vivo coordinated regulations of the FGF23-α-Klotho system is limited and requires additional carefully designed studies.

In summary, our pharmacological and genetic manipulation studies demonstrate that the phosphate-lowering effect of FGF23 is a WNT-independent process, particularly in *Hyp* mice lacking *PHEX* activities. Additional studies are needed to determine the signaling network involved in the regulation of FGF23-mediated phosphate metabolism. The *Hyp/Lrp6* double mutant mice we have generated provide a unique in vivo tool to study WNT-independent downstream signaling events exerted by FGF23. Detailed studies to determine the molecular interactions among FGF23, FGF receptor and αKlotho, and downstream signaling targets will provide important insights into the physiologic regulation of phosphate metabolism and pathologic complications, arising from its abnormal regulation [43–45].

Acknowledgements

Part of the research is supported by NIH grants R01-DK077276 to M.S.R. and R01-DE13686 to M.M., and the Japan Society for the Promotion of Science (JSPS), Excellent Young Researchers Overseas Visit Program grant (21-8134) to K.U.

References

1 Gaasbeek A, Meinders AE: Hypophosphatemia: an update on its etiology and treatment. Am J Med 2005;118:1094–1101.

2 Razzaque MS: Phosphate toxicity: new insights into an old problem. Clin Sci (Lond) 2011;120:91–97.

3 Razzaque MS: Osteo-renal regulation of systemic phosphate metabolism. IUBMB Life 2011;63:240–247.

4 Drezner MK: Hypophosphatemic rickets. Endocr Dev 2003;6:126–155.

5 Osuka S, Razzaque MS: Can features of phosphate toxicity appear in normophosphatemia? J Bone Miner Res 2012;30:10–18.

6 Murer H, Hernando N, Forster I, Biber J: Proximal tubular phosphate reabsorption: molecular mechanisms. Physiol Rev 2000;80: 1373–1409.

7 Razzaque MS: Therapeutic potential of klotho-fgf23 fusion polypeptides: Wo2009095372. Expert Opin Ther Pat 2010; 20:981–985.

8 Razzaque MS: The fgf23-klotho axis: endocrine regulation of phosphate homeostasis. Nat Rev Endocrinol 2009;5:611–619.

9 Razzaque MS: Fgf23-mediated regulation of systemic phosphate homeostasis: Is klotho an essential player? Am J Physiol Renal Physiol 2009;296:F470–F476.

10 Razzaque MS: Does fgf23 toxicity influence the outcome of chronic kidney disease? Nephrol Dial Transplant 2009;24:4–7.

11 Razzaque MS, Lanske B: The emerging role of the fibroblast growth factor-23-klotho axis in renal regulation of phosphate homeostasis. J Endocrinol 2007;194:1–10.

12 Razzaque MS: Can fibroblast growth factor 23 fine-tune therapies for diseases of abnormal mineral ion metabolism? Nat Clin Pract Endocrinol Metab 2007;3:788–789.

13 Liu S, Tang W, Fang J, Ren J, Li H, Xiao Z, Quarles LD: Novel regulators of fgf23 expression and mineralization in hyp bone. Mol Endocrinol 2009;23:1505–1518.

14 Farrow EG, Summers LJ, Schiavi SC, McCormick JA, Ellison DH, White KE: Altered renal fgf23-mediated activity involving mapk and wnt. Effects of the hyp mutation. J Endocrinol 2010;207:67–75.

15 Andersson ER, Bryjova L, Biris K, Yamaguchi TP, Arenas E, Bryja V: Genetic interaction between lrp6 and wnt5a during mouse development. Dev Dyn 2010;239:237–245.

16 Zhou CJ, Molotkov A, Song L, Li Y, Pleasure DE, Pleasure SJ, Wang YZ: Ocular coloboma and dorsoventral neuroretinal patterning defects in lrp6 mutant eyes. Dev Dyn 2008; 237:3681–3689.

17 Lindvall C, Zylstra CR, Evans N, West RA, Dykema K, Furge KA, Williams BO: The wnt co-receptor lrp6 is required for normal mouse mammary gland development. PLoS One 2009;4:e5813.

18 Glinka A, Wu W, Delius H, Monaghan AP, Blumenstock C, Niehrs C: Dickkopf-1 is a member of a new family of secreted proteins and functions in head induction. Nature 1998;391:357–362.

19 Mao J, Wang J, Liu B, Pan W, Farr GH 3rd, Flynn C, Yuan H, Takada S, Kimelman D, Li L, Wu D: Low-density lipoprotein receptor-related protein-5 binds to axin and regulates the canonical wnt signaling pathway. Mol Cell 2001;7:801–809.

20 Semenov MV, Tamai K, Brott BK, Kuhl M, Sokol S, He X: Head inducer dickkopf-1 is a ligand for wnt coreceptor lrp6. Curr Biol 2001;11:951–961.

21 Bafico A, Liu G, Yaniv A, Gazit A, Aaronson SA: Novel mechanism of wnt signalling inhibition mediated by dickkopf-1 interaction with lrp6/arrow. Nat Cell Biol 2001;3:683–686.

22 Holmen SL, Giambernardi TA, Zylstra CR, Buckner-Berghuis BD, Resau JH, Hess JF, Glatt V, Bouxsein ML, Ai M, Warman ML, Williams BO: Decreased bmd and limb deformities in mice carrying mutations in both lrp5 and lrp6. J Bone Miner Res 2004;19: 2033–2040.

23 Kokubu C, Heinzmann U, Kokubu T, Sakai N, Kubota T, Kawai M, Wahl MB, Galceran J, Grosschedl R, Ozono K, Imai K: Skeletal defects in ringelschwanz mutant mice reveal that lrp6 is required for proper somitogenesis and osteogenesis. Development 2004;131: 5469–5480.

Razzaque MS (ed): Phosphate and Vitamin D in Chronic Kidney Disease.
Contrib Nephrol. Basel, Karger, 2013, vol 180, pp 138–148 (DOI: 10.1159/000346793)

Can Salivary Phosphate Levels Be an Early Biomarker to Monitor the Evolvement of Obesity?

Mor-Li Hartman[a] · Francisco Groppo[b] · Mutsuko Ohnishi[c] · J. Max Goodson[a] · Hatice Hasturk[a] · Mary Tavares[a] · Tina Yaskell[a] · Constantino Floros[a] · Kazem Behbehani[d] · Mohammed S. Razzaque[c]

[a]Department of Applied Oral Sciences, The Forsyth Institute, Cambridge, Mass., and [b]Piracicaba Dental School, State University of Campinas, Piracicaba, SP, Brazil; [c]Department of Oral Medicine, Infection and Immunity, Harvard School of Dental Medicine, Boston, Mass., USA; [d]The Dasman Diabetes Institute, Dasman, Kuwait

Abstract

Phosphate is an essential nutrient required for important biological reactions that maintain the normal homoeostatic control of the cell. The adverse effects of phosphate metabolism in obesity have not been studied in detail, chiefly because such an association is thought to be uncommon. However, in some animal models of obesity, serum phosphate levels were noted to be higher than the nonobese controls. For example, leptin-deficient (ob/ob) mice become severely obese and have high serum phosphate levels. In this study, we analyzed the phosphate content in saliva collected from children ($n = 77$; 10.5 ± 1.8) to evaluate association with body mass index; there is a significant increase of salivary phosphate content in obese compared to normal-weight children (ANOVA $p < 0.001$). The correlation coefficient (r) between BMI and phosphate was 0.33 ($p = 0.0032$). Our results suggest that the human salivary phosphate level may be an early biomarker of the genesis of obesity in children. The diagnostic importance lies in the fact that the salivary phosphate level could provide a noninvasive predictive marker in the development of obesity. Further studies will be required to understand the underlying mechanism of increased salivary phosphate accumulation in obese and overweight children. Nevertheless, its occurrence without systemic changes could be of diagnostic value, particularly in monitoring evolvement of obesity.

The increasing occurrence of obesity and its related complications, including type 2 diabetes mellitus and cardiovascular anomalies, is alarming and becoming a major public health problem. Despite significant improvement of our understanding of underlying molecular mechanisms of adipose tissue remodeling and eventual development of obesity, potential factors that might serve as a biomarker to monitor evolvement of obesity is not yet clearly identified. Earlier studies have found that the composition of salivary bacteria changes in overweight women [1]. In particular, the percentage of the bacterium *Selenomonas noxia* was capable of identifying 98.4% of studied 313 overweight women from a group of generally healthy individuals, with a remarkably high degree of diagnostic power [1]. Whether change in the composition of salivary content could serve as biological marker of a developing overweight condition is an area that needs further studies. Of even greater interest, and the subject of future research, is the possibility of using easily available saliva as a diagnostic and prognostic marker of systemic diseases, including obesity.

A clinical association between phosphate imbalance and metabolic diseases has been reported elsewhere. For instance, serum phosphate levels were positively correlated with insulin sensitivity (but not with insulin secretion) in a study conducted in 881 individuals without diabetes mellitus [2]; such a correlation was independent of age, sex, proportion of body adipose content, serum calcium and serum creatinine levels [2]. In a separate study of 298 children and adolescents aged 6–12 years old, in which 190 individuals with obesity and 108 controls without obesity were compared, reduced phosphate serum levels were significantly associated with the development of insulin resistance in children with obesity [3]. Moreover, in a leptin-deficient animal model of obesity, higher serum phosphate levels were noted, as compared to the nonobese controls [4–6]. Summarizing the above-mentioned evidence, it appears likely there might be an association between phosphate metabolism and adipose tissue turnover. This study was designed to study whether the phosphate content of saliva might be clinically useful in monitoring the development of obesity.

The physiological phosphate balance is mostly accomplished by cross-organ talk among kidney, intestine, and bone, where sodium-dependent phosphate transporters (Na/Pi) play a key role in the absorption and reabsorption of consumed phosphate [7–12]. The Na/Pi-2b is mostly present in the luminal side of the intestinal cells where it helps in intestinal phosphate absorption according to the body's need of the phosphate [13]. Almost 80% of the filtered phosphate in the kidney is reabsorbed in the proximal tubular epithelial cells and is also partly accomplished by sodium-dependent phosphate uptake through Na/Pi-2a and Na/Pi-2c transporters. Of relevance, Na/Pi-2b transporters are also expressed in the salivary gland; studies have identified the expression of Na/Pi-2b mRNA and protein in the salivary duct cells [14, 15]. In fact, the ductal cells mostly demonstrated apical Na/Pi-2b expression in salivary glands, whereas the acinar cells showed mainly basolateral expression, suggesting that there might be phosphate secretion by the acini and phosphate reabsorption by the ducts [15]. Moreover, klotho, a cofactor for circulating fibroblast growth factor 23 (FGF23), a master regulator of phosphate metabolism [11, 16–18], is also present in the salivary gland [19], and is therefore providing the necessary machinery for a local salivary regulation of phosphate metabolism, dependent or independent of systemic regulation. This study provides the data to show that the phosphate content in saliva collected from children has a close association with BMI.

Materials and Methods

Saliva and Blood Collection from Children

Saliva and blood samples were collected from 77 children (59% male), aged 10.5 ± 1.8 years, under fasting conditions. Height, weight, blood pressure, heart rate and fitness were measured and recorded. Fitness was measured by heart rate elevation to a Queens College step test [20]. BMI percentile was used to define the children into body weight categories. This study was reviewed and approved by the Forsyth Institute institutional review board. Informed consent was obtained from parents or guardians and assent was obtained from each child. The study was conducted in Portland, Me., USA.

Measurement of Phosphate Contents

Paired 100-μl samples of saliva and plasma from each child were prepared and maintained frozen at –80°C until assayed. Phosphate was measured on a nontargeted metabolic profiling platform (Metabolon, Durham, N.C., USA) employing gas chromatography/mass spectrometry (GC/MS) [21, 22]. For each sample, 100 μl of plasma or saliva was used for analyses. Using an automated liquid handler (Hamilton LabStar, Salt Lake City, Utah, USA), protein was precipitated with methanol that contained four standards to report on extraction efficiency. The resulting supernatant was split into equal aliquots for analysis. One aliquot, dried under nitrogen and vacuum-desiccated, was subsequently derivatized to a final volume of 50 μl for GC/MS analysis using equal parts of bis(trimethylsilyl)trifluoroacetamide and the solvent mixtureacetonitrile:dichloromethane:cyclohexane (5:4:1) with 5% triethylamine at 60°C for 1 h. Mass spectrometric salivary phosphate data were tested and found non-normal using the Shapiro-Wilk statistics and were also both skewed and kurtotic. Saliva phosphate levels were also determined by colorimetric measurements using the Stanbio Phosphorus Liqui-UV Test (Stanbio Laboratory, Boerne, Tex., USA) as detailed earlier [18, 23, 24]. Statistical evaluations of these data were performed by the Mann-Whitney test. BMI percentile was computed from software from the Center for Disease Control [25]. Body weight categories were assigned using their percentile recommendations (normal healthy weight <85%, overweight ≥85–95% and obese ≥95%). Two children were classified as underweight (percentile <5%). These were included in the normal healthy weight group for analysis.

Leptin-Deficient Obese Mice
Heterozygous leptin-deficient obese [C57Bl/6J lep^{ob} (+/–)] mutants were obtained from Jackson Laboratory, Me., USA and maintained in accordance with the NIH Guide for the Care and Use of Laboratory Animals. Mouse breeding was performed according to the protocols approved by the institution's subcommittee on animal care (IACUC). Routine PCR using genomic DNA extracted from tail clips was performed for genotyping of the various groups of mice, as detailed earlier [23, 26, 27]. The total body weight of wild-type and *ob/ob* mice were taken weekly starting at 3 weeks of age until 30 weeks. At least three or more mice from each genotype were sacrificed at 9 weeks and retroperitoneal, mesenteric and epididymal fat tissue weights were recorded. In addition, liver weights of the wild-type and *ob/ob* mice were taken before fixing part of the liver for histological analysis. Fasting blood glucose levels were determined in all four genotypes. In addition, serum isolated from blood obtained by cheek-pouch bleeding of wild-type and *ob/ob* mice was isolated and stored at –80 °C. Serum phosphate levels were determined by colorimetric measurements using the Stanbio Phosphorus Liqui-UV Test [26, 28, 29]. Statistically significant differences between groups were evaluated either by the Student's t test or the Mann-Whitney U test for a comparison between two groups. All values are expressed as mean ± SE. $p < 0.05$ was considered to be statistically significant.

Results

Animal Studies

Leptin-deficient obese (ob/ob) mice develop hyperglycemia, glucose intolerance, and obesity. Since the ob/ob mouse could not produce leptin, its food intake is uncontrolled due to the lack of feedback signaling [30]. Mutant ob/ob mice are indistinguishable from their unaffected wild-type littermates at birth, but gain weight rapidly throughout their lives, reaching a weight three times that of unaffected wild-type mice (fig. 1, upper left panel). Compared to the wild-type controls, the increased body size of ob/ob mice is also associated with significantly increased fat accumulation, as determined by analysis of retroperitoneal, mesenteric and epididymal fat tissue weights. Increased accumulation of fat is also obvious in the livers obtained from ob/ob mice (fig. 1, right panel). The ob/ob mice also develop hyperglycemia [31] with increased serum levels of phosphate (fig. 1, lower left panel). Although increased serum levels of phosphate in ob/ob mice could partly be related to the increased food consumption, it is, however, an intriguing association that needs further exploration.

Human Study

Biometric characteristics of the 77 children admitted into the study are shown in table 1. Differences between normal and obese children (p 1:3) were all statistically significant except for diastolic blood pressure and heart rate. The ac-

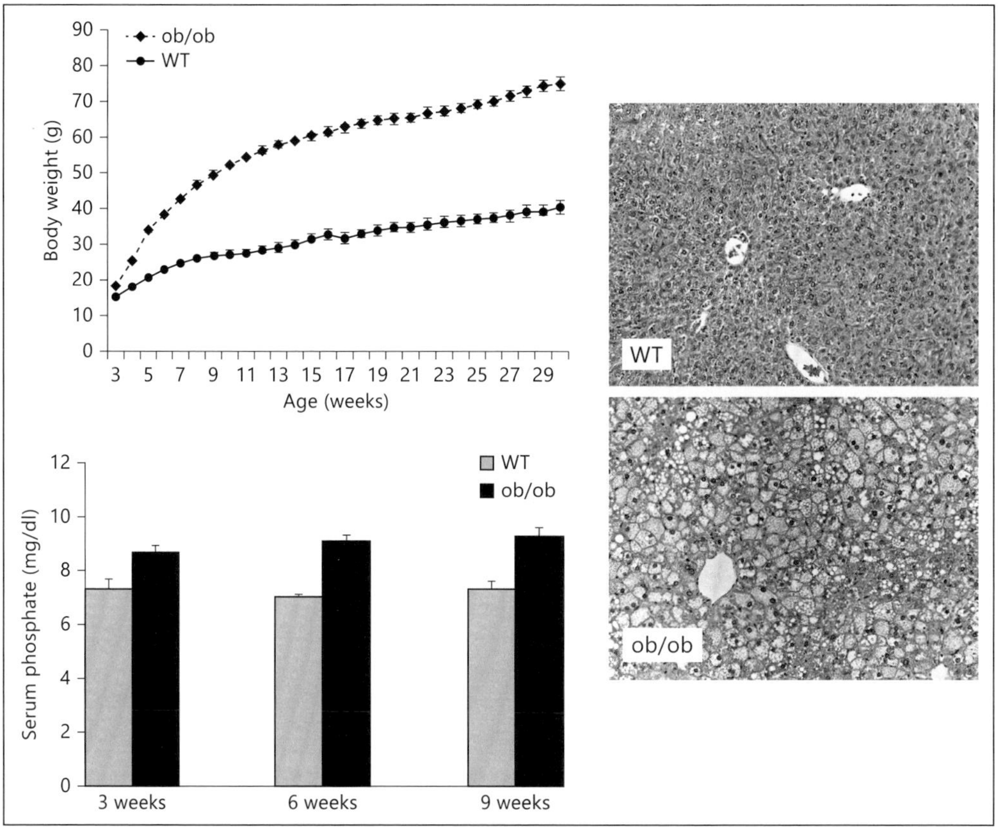

Fig. 1. Upper left panel: The body weight chart of wild-type and ob/ob animals. Note that compared to the wild-type control animals, the ob/ob mice are significantly heavier due to an excessive accumulation of fat tissues. Right panel: In accord with the relatively increased liver weight in ob/ob mice over wild-type controls, the histological analysis of the liver shows increased fat tissue accumulation in ob/ob mice. Lower left panel: Biochemical analysis of serum phosphate levels. Note that the serum phosphate levels are significantly higher in ob/ob mice compared with the wild-type (WT) mice at 3, 6 and 9 weeks of age. * $p < 0.05$ vs. WT. Serum samples collected from at least 4 or more mice were used for the measurements.

ceptance criteria (10 ± 1 year) was expanded (10 ± 2 years) for overweight and obese volunteers to obtain a sufficient number willing to provide both saliva and blood samples. This resulted in small but significant differences in age for heavier weight groups. The correlation and statistical significance between salivary phosphate and other parameters are presented in table 2. We have found that different obesity parameters, including BMI, waist circumference, fitness and weight, along with cardiovascular parameter (systolic blood pressure) were significantly associated with salivary phosphate content. Saliva determination

Table 1. Characteristics of children investigated in this study

	Normal weight (n = 36)			Overweight (n = 15)			Obese (n = 26)		
	median	intraquartile range	p = 1:3	median	intraquartile range	p = 1:2	median	intraquartile range	p = 2:3
Weight, lb	74.4	24.4	<0.0001	106.1	22.4	<0.0001	151.5	50.1	0.002
Height, inches	56.5	5.0	0.004	58.5	4.9	0.07	60.1	9.5	0.3
Age, years	9.6	2.4	0.02	10.5	1.9	0.02	10.8	2.5	0.8
BMI	17.2	2.2	<0.0001	21.1	2.4	<0.0001	27.4	7.6	<0.0001
Waist, inches	24.0	3.5	<0.0001	27.5	5.0	<0.0001	34.8	9.0	0.004
Systolic BP, mm Hg	117.5	15.2	0.02	122.0	12.0	0.03	121.2	14.7	0.9
Diastolic BP, mm Hg	67.3	8.5	0.1	69.0	6.7	0.2	71.2	8.7	0.9
Heart rate, BPM	77.3	15.5	0.7	73.3	23.3	0.3	81.8	20.3	0.6
Fitness, BPM	24.7	16.7	0.002	37.2	25.5	0.02	35.1	15.3	0.9
Salivary flow, ml/h	37.3	27.4	0.1	40.3	25.3	0.9	32.7	27.5	0.3

Values listed include median, intraquartile range and the significance p value of differences between body weight categories (1 = normal, 2 = overweight, 3 = obese) by Mann-Whitney analysis.

Table 2. Correlation (r) and statistical significance (p) between salivary phosphate and each measure

Measure	r	p
BMI	0.33	0.003
Waist, inches	0.31	0.007
Fitness, BPM	0.29	0.011
Weight, lb	0.27	0.018
Systolic BP, mm Hg	0.22	0.051
Heart rate, BPM	0.21	0.067
Diastolic BP, mm Hg	0.17	0.139
Height, inches	0.09	0.433
Age, years	0.01	0.913
Salivary flow, ml/h	–0.24	0.035

Results indicate that measures of obesity (BMI, waist circumference, fitness and weight), one cardiovascular parameter (systolic blood pressure) and salivary flow were significantly associated with salivary phosphate.

by nontargeted mass spectrometric analysis revealed that phosphate was significantly higher in the saliva of obese children. By comparison, however, plasma phosphate was not associated with obesity (fig. 2), suggesting that changes in salivary phosphate levels could occur even before serum changes become apparent, and thereby could be used as a biomarker for monitoring evolvement

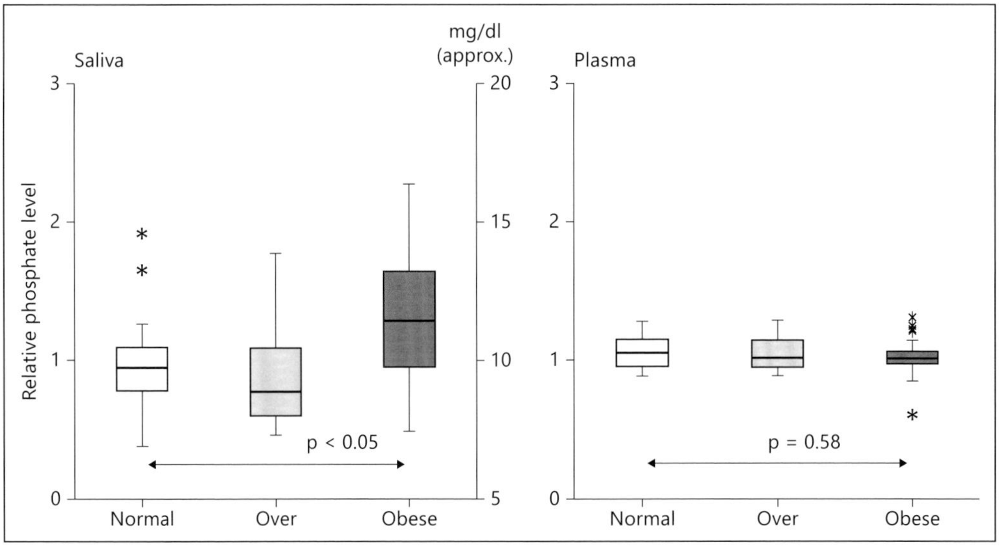

Fig. 2. Relative phosphate levels in saliva and plasma by mass spectroscopy. Central bars are the median, rectangles are the intraquartile range, whiskers are ×1.5 intraquartile range, asterisks are outside values and p is computed using the Mann-Whitney test. Relative phosphate values were obtained by dividing the spectrograph response by the median of all values. This comparison indicates that salivary phosphate levels in obese children were significantly greater than in normal healthy weight children whereas plasma levels were not. Approximate concentrations (mg/dl) were computed from colorometric determinations of aliquots.

of obesity. Of relevance, an r^2 of 0.70 and a p of 0.003 for prediction of phosphate from the spectrographic response by the quantitative phosphate levels (mg/dl) = 5.03 MSR + 4.92. Thus, the median mass spectrographic value for saliva of 1.0 would roughly equivalent to 9.95 mg/dl (fig. 3).

Discussion

Child obesity has accelerated over the last decade [32]. As a result, investigation of biomarkers that can serve as early warning signs has become increasingly important. Salivary analysis ranks high in consideration due to ease of collection in a population of children that are understandably reticent to provide blood samples. In this report, we provide evidence that salivary phosphate is elevated in obese children. By contrast, plasma phosphate was not found to be elevated in these children. Taken together, these observations suggest that phosphate metabolism might be associated with fat cell turnover, and that

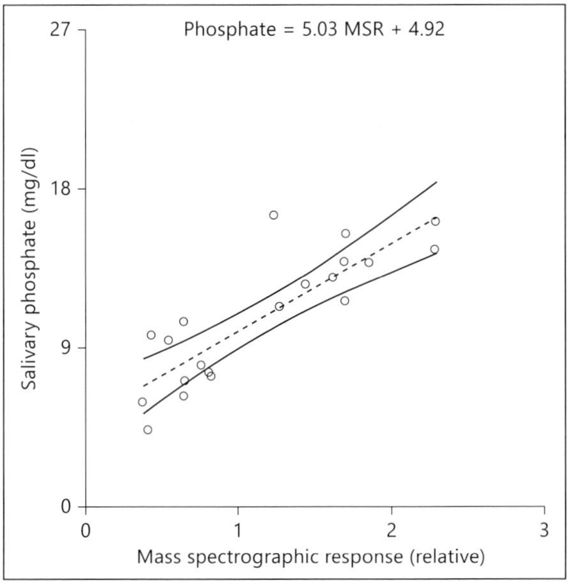

Fig. 3. Least-squares fit linear function and 95% CIs. The relationship has an r^2 of 0.72 and a $p < 0.0001$ for prediction of phosphate from the spectrographic response by the function phosphate (mg/dl) = 5.03 MSR + 4.92. Thus, the median mass spectrographic value for saliva of 1.0 would be 9.95 mg/dl.

salivary phosphate could be an early marker of metabolic disease associated with development of obesity.

In this study, we presented values for phosphate that were obtained by colorimetric phosphate test from the supernatant of unstimulated saliva (fig. 2). The similar phosphate values in our study are in the similar range compared to other published studies; for instance, studies that were done on healthy children presenting a range of salivary phosphate concentration of 10.5–17 mg/dl [33–35]. No data on BMI were presented in these published studies, except for one study that stated that children with metabolic disorder were excluded [34]. It is, therefore, not possible to estimate the accurate phosphate concentration from these studies, as some children might have been overweight. In a similar line of study that was done on children, adolescents and young adults who have chronic kidney disease showed salivary phosphate values for the healthy subjects of 15.8 mg/dl and patients (on predialytic) of 20.77 mg/dl [36].

Why salivary phosphate levels are high in obese children is not yet clear, and might be related to the altered functions of phosphate-regulating machineries present in the salivary glands. It is important to note that type II sodium-phosphate cotransporters (Na/Pi-2b) are present in the salivary gland [14, 15]. In addition, the type III sodium phosphate cotransporters, PiT1 and PiT2, have been detected in human salivary glands [37]. The possible role of PiT1 and PiT2 in the salivary gland has not been well studied, but a number of studies demonstrated the role for both PiT1 and PiT2 in phosphate reabsorption in the kidney. Depending on the sodium phosphate cotransporter activities in the salivary gland, it is likely that there might be a local regulation of phosphate metabolism that is independent of systemic phosphate regulation. In fact, such possibilities are raised from our observations of increased salivary phosphate contents in obese children, despite their normal plasma phosphate levels. It is important to note that despite recognition that serum phosphate level is the gold standard to estimate the overall phosphate status of the body, the amount of intracellular phosphate or phosphate storage is not taken into consideration in such traditional methods of phosphate measurement. Both human and experimental studies have shown clearly that certain features of phosphate toxicity might appear even in normophosphatemic conditions [38], thereby exposing the limitation of serum phosphate measurements to detect early events of phosphate toxicity. In that context, our result of increased salivary phosphate content without any change in plasma levels provides a meaningful early marker for eventual obesity development.

Since consumption of soda drinks and fast foods are well connected with childhood obesity [39], it will be interesting to know whether such phosphate-rich foods and drinks, by local regulation, might increase the salivary phosphate content that we found in our obese cohorts. Moreover, obese children exhibit more bone resorption than healthy normal-weight children [40], and thereby could induce altered mineral ion metabolism. Further studies will explain the underlying mechanism of increased salivary phosphate accumulation in obese and overweight children, with its biological and clinical significances. This particular study, although limited in numbers of subjects involved, provides the evidence of potential clinical importance of salivary phosphate in monitoring the evolvement of obesity.

Acknowledgements

This work was funded by the grants from the Dasman Institute, Kuwait (to J.M.G), the US National Institutes of Health grant No. DK077276 (to M.S.R.), and the institutional support from the Harvard School of Dental Medicine (to M.S.R).

References

1 Goodson JM, Groppo D, Halem S, Carpino E: Is obesity an oral bacterial disease? J Dent Res 2009;88:519–523.

2 Haap M, Heller E, Thamer C, Tschritter O, Stefan N, Fritsche A: Association of serum phosphate levels with glucose tolerance, insulin sensitivity and insulin secretion in non-diabetic subjects. Eur J Clin Nutr 2006;60: 734–739.

3 Celik N, Andiran N: The relationship between serum phosphate levels with childhood obesity and insulin resistance. J Pediatr Endocrinol Metab 2011;24:81–83.

4 Razzaque MS: The role of klotho in energy metabolism. Nat Rev Endocrinol 2012;8:579–587.

5 Tsuji K, Maeda T, Kawane T, Matsunuma A, Horiuchi N: Leptin stimulates fibroblast growth factor 23 expression in bone and suppresses renal 1alpha,25-dihydroxyvitamin d3 synthesis in leptin-deficient mice. J Bone Miner Res 2010;25:1711–1723.

6 Ohnishi M, Kato S, Akiyoshi J, Atfi A, Razzaque MS: Dietary and genetic evidence for enhancing glucose metabolism and reducing obesity by inhibiting klotho functions. FASEB J 2011;25:2031–2039.

7 Razzaque MS: Phosphate toxicity: new insights into an old problem. Clin Sci (Lond) 2011;120:91–97.

8 Razzaque MS: Osteo-renal regulation of systemic phosphate metabolism. IUBMB Life 2011;63:240–247.

9 Gattineni J, Baum M: Regulation of phosphate transport by fibroblast growth factor 23 (fgf23): implications for disorders of phosphate metabolism. Pediatr Nephrol 2010;25: 591–601.

10 Razzaque MS: Fgf23-mediated regulation of systemic phosphate homeostasis: Is klotho an essential player? Am J Physiol Renal Physiol 2009;296:F470–F476.

11 Razzaque MS: The fgf23-klotho axis: endocrine regulation of phosphate homeostasis. Nat Rev Endocrinol 2009;5:611–619.

12 Lu Y, Feng JQ: Fgf23 in skeletal modeling and remodeling. Curr Osteoporos Rep 2011; 9:103–108.

13 Tenenhouse HS: Regulation of phosphorus homeostasis by the type IIa Na/phosphate cotransporter. Annu Rev Nutr 2005;25:197–214.

14 Homann V, Rosin-Steiner S, Stratmann T, Arnold WH, Gaengler P, Kinne RK: Sodium-phosphate cotransporter in human salivary glands: molecular evidence for the involvement of NPT2b in acinar phosphate secretion and ductal phosphate reabsorption. Arch Oral Biol 2005;50:759–768.

15 Lederer E, Miyamoto K: Clinical consequences of mutations in sodium phosphate cotransporters. Clin J Am Soc Nephrol 2012;7:1179–1187.

16 Goetz R, Ohnishi M, Kir S, Kurosu H, Wang L, Pastor J, Ma J, Gai W, Kuro-o M, Razzaque MS, Mohammadi M: Conversion of a paracrine fibroblast growth factor into an endocrine fibroblast growth factor. J Biol Chem 2012;287:29134–29146.

17 Goetz R, Ohnishi M, Ding X, Kurosu H, Wang L, Akiyoshi J, Ma J, Gai W, Sidis Y, Pitteloud N, Kuro OM, Razzaque MS, Mohammadi M: Klotho coreceptors inhibit signaling by paracrine fibroblast growth factor 8 subfamily ligands. Mol Cell Biol 2012;32: 1944–1954.

18 Nakatani T, Sarraj B, Ohnishi M, Densmore MJ, Taguchi T, Goetz R, Mohammadi M, Lanske B, Razzaque MS: In vivo genetic evidence for klotho-dependent, fibroblast growth factor 23 (FGF23)-mediated regulation of systemic phosphate homeostasis. FASEB J 2009;23:433–441.

19 Amano I, Imaizumi Y, Kaji C, Kojima H, Sawa Y: Expression of podoplanin and classical cadherins in salivary gland epithelial cells of klotho-deficient mice. Acta Histochem Cytochem 2011;44:267–276.

20 Suriano K, Curran J, Byrne SM, Jones TW, Davis EA: Fatness, fitness, and increased cardiovascular risk in young children. J Pediatr 2010;157:552–558.

21 Evans AM, DeHaven CD, Barrett T, Mitchell M, Milgram E: Integrated, nontargeted ultrahigh performance liquid chromatography/ electrospray ionization tandem mass spectrometry platform for the identification and relative quantification of the small-molecule complement of biological systems. Anal Chem 2009;81:6656–6667.

22 Ohta T, Masutomi N, Tsutsui N, Sakairi T, Mitchell M, Milburn MV, Ryals JA, Beebe KD, Guo L: Untargeted metabolomic profiling as an evaluative tool of fenofibrate-induced toxicology in Fischer 344 male rats. Toxicol Pathol 2009;37:521–535.
23 Nakatani T, Ohnishi M, Razzaque MS: Inactivation of klotho function induces hyperphosphatemia even in presence of high serum fibroblast growth factor 23 levels in a genetically engineered hypophosphatemic (Hyp) mouse model. FASEB J 2009;23:3702–3711.
24 Ohnishi M, Kato S, Razzaque MS: Genetic induction of phosphate toxicity significantly reduces the survival of hypercholesterolemic obese mice. Biochem Biophys Res Commun 2011;415:434–438.
25 CDC: Child and teen BMI calculator. http://apps.nccd.cdc.gov/dnpabmi/, 2012.
26 Ohnishi M, Razzaque MS: Dietary and genetic evidence for phosphate toxicity accelerating mammalian aging. FASEB J 2010;24:3562–3571.
27 Nakatani T, Bara S, Ohnishi M, Densmore MJ, Taguchi T, Goetz R, Mohammadi M, Lanske B, Razzaque MS: In vivo genetic evidence of klotho-dependent functions of fgf23 in regulation of systemic phosphate homeostasis. FASEB J 2009;23:433–441.
28 Ohnishi M, Nakatani T, Lanske B, Razzaque MS: In vivo genetic evidence for suppressing vascular and soft-tissue calcification through the reduction of serum phosphate levels, even in the presence of high serum calcium and 1,25-dihydroxyvitamin d levels. Circ Cardiovasc Genet 2009;2:583–590.
29 Ohnishi M, Nakatani T, Lanske B, Razzaque MS: Reversal of mineral ion homeostasis and soft-tissue calcification of klotho knockout mice by deletion of vitamin d 1alpha-hydroxylase. Kidney Int 2009;75:1166–1172.
30 Friedman JM, Halaas JL: Leptin and the regulation of body weight in mammals. Nature 1998;395:763–770.
31 Kamohara S, Burcelin R, Halaas JL, Friedman JM, Charron MJ: Acute stimulation of glucose metabolism in mice by leptin treatment. Nature 1997;389:374–377.
32 Hsia DS, Fallon SC, Brandt ML: Adolescent bariatric surgery. Arch Pediatr Adolesc Med 2012;166:757–766.
33 Damle SG VI, Yadav R, Bhattal H, Loomba A: Quantitative determination of inorganic constituents in saliva and their relationship with dental caries experience in children. Dentistry 2012;2:131.
34 Shahrabi M, Nikfarjam J, Alikhani A, Akhoundi N, Ashtiani M, Seraj B: A comparison of salivary calcium, phosphate, and alkaline phosphatase in children with severe, moderate caries, and caries free in Tehran's kindergartens. J Indian Soc Pedod Prev Dent 2008;26:74–77.
35 Wu KP, Ke JY, Chung CY, Chen CL, Hwang TL, Chou MY, Wong AM, Hu CF, Lee YC: Relationship between unstimulated salivary flow rate and saliva composition of healthy children in Taiwan. Chang Gung Med J 2008;31:281–286.
36 Davidovich E, Davidovits M, Peretz B, Shapira J, Aframian DJ: The correlation between dental calculus and disturbed mineral metabolism in paediatric patients with chronic kidney disease. Nephrol Dial Transplant 2009;24:2439–2445.
37 Corut A, Senyigit A, Ugur SA, Altin S, Ozcelik U, Calisir H, Yildirim Z, Gocmen A, Tolun A: Mutations in slc34a2 cause pulmonary alveolar microlithiasis and are possibly associated with testicular microlithiasis. Am J Hum Genet 2006;79:650–656.
38 Osuka S, Razzaque MS: Can features of phosphate toxicity appear in normophosphatemia? J Bone Miner Metab 2012;30:10–18.
39 Dhingra R, Sullivan L, Jacques PF, Wang TJ, Fox CS, Meigs JB, D'Agostino RB, Gaziano JM, Vasan RS: Soft drink consumption and risk of developing cardiometabolic risk factors and the metabolic syndrome in middle-aged adults in the community. Circulation 2007;116:480–488.
40 Dimitri P, Wales JK, Bishop N: Adipokines, bone-derived factors and bone turnover in obese children: evidence for altered fat-bone signalling resulting in reduced bone mass. Bone 2011;48:189–196.

J. Max Goodson, DDS, PhD
Department of Applied Oral Sciences, The Forsyth Institute
245 First Street
Cambridge MA 02142 (USA)
E-Mail Mgoodson@forsyth.org

Author Index

Subject Index